STATUTORY INSTRUM

1999 No. 3232

HEALTH AND SAFETY

The Ionising Radiations Regulations 1999

Made - - - -	*3rd December 1999*
Laid before Parliament	*9th December 1999*
Coming into force	
All regulations except for regulation 5 -	*1st January 2000*
Regulation 5 - - -	*13th May 2000*

ARRANGEMENT OF REGULATIONS

PART I

INTERPRETATION AND GENERAL

1. Citation and commencement.
2. Interpretation.
3. Application.
4. Duties under the Regulations.

PART II

GENERAL PRINCIPLES AND PROCEDURES

5. Authorisation of specified practices.
6. Notification of specified work.
7. Prior risk assessment etc.
8. Restriction of exposure.
9. Personal protective equipment.
10. Maintenance and examination of engineering controls etc. and personal protective equipment.
11. Dose limitation.
12. Contingency plans.

[DETR 1686]

PART III

ARRANGEMENTS FOR THE MANAGEMENT OF RADIATION PROTECTION

13. Radiation protection adviser.
14. Information, instruction and training.
15. Co-operation between employers.

PART IV

DESIGNATED AREAS

16. Designation of controlled or supervised areas.
17. Local rules and radiation protection supervisors.
18. Additional requirements for designated areas.
19. Monitoring of designated areas.

PART V

CLASSIFICATION AND MONITORING OF PERSONS

20. Designation of classified persons.
21. Dose assessment and recording.
22. Estimated doses and special entries.
23. Dosimetry for accidents etc.
24. Medical surveillance.
25. Investigation and notification of overexposure.
26. Dose limitation for overexposed employees.

PART VI

ARRANGEMENTS FOR THE CONTROL OF RADIOACTIVE SUBSTANCES, ARTICLES AND EQUIPMENT

27. Sealed sources and articles containing or embodying radioactive substances.
28. Accounting for radioactive substances.
29. Keeping and moving of radioactive substances.
30. Notification of certain occurrences.
31. Duties of manufacturers etc. of articles for use in work with ionising radiation.
32. Equipment used for medical exposure.
33. Misuse of or interference with sources of ionising radiation.

PART VII

DUTIES OF EMPLOYEES AND MISCELLANEOUS

34. Duties of employees.
35. Approval of dosimetry services.
36. Defence on contravention.
37. Exemption certificates.
38. Extension outside Great Britain.
39. Transitional provisions.
40. Modifications relating to the Ministry of Defence.
41. Modification, revocation and saving.

Schedule 1. Work not required to be notified under regulation 6.
Schedule 2. Particulars to be provided in a notification under regulation 6(2).
Schedule 3. Additional particulars that the Executive may require.
Schedule 4. Dose limits.
Schedule 5. Matters in respect of which a radiation protection adviser must be consulted by a radiation employer.
Schedule 6. Particulars to be entered in the radiation passbook.
Schedule 7. Particulars to be contained in a health record.
Schedule 8. Quantities and concentrations of radionuclides.
Schedule 9. Modifications.

The Secretary of State, being the designated(a) Minister for the purpose of section 2(2) of the European Communities Act 1972(b) in relation to measures relating to the basic safety standards for the protection of the general public and workers against the dangers of ionising radiation, in exercise of the powers conferred by the said section 2(2) and sections 15(1), (2), (3)(a), (4)(a), (5)(b), (6)(b) and (9), 43(2), (4), (5) and (6), 52(2) and (3), 80(1) and 82(3)(a) of, and paragraphs 1(1) and (2), 3, 6, 7, 8, 9, 11, 13, 14, 15(1), 16, 20 and 21(a) and (b) of Schedule 3 to, the Health and Safety at Work etc. Act 1974(c) ("the 1974 Act") and of all the powers enabling him in that behalf—

(a) for the purpose of giving effect without modifications to proposals submitted to him by the Health and Safety Commission under section 11(2)(d) of the 1974 Act after the carrying out by the Commission of consultations in accordance with section 50(3) of that Act; and

(b) it appearing to him that the modifications set out in paragraphs 1 and 2 of Schedule 9 to the Regulations are expedient and that it also appearing to him not to be appropriate to consult bodies in respect of such modifications in accordance with section 80(4) of the 1974 Act,

hereby makes the following Regulations:—

PART I

INTERPRETATION AND GENERAL

Citation and commencement

1. These Regulations may be cited as the Ionising Radiations Regulations 1999 and shall come into force—

(a) as respect all regulations except for regulation 5, on 1st January 2000; and

(b) as respects regulation 5, on the 13th May 2000.

Interpretation

2.—(1) In these Regulations, unless the context otherwise requires—

"accelerator" means an apparatus or installation in which particles are accelerated and which emits ionising radiation with an energy higher than 1MeV;

"appointed doctor" means, subject to regulation 39(5) (which relates to transitional provisions), a registered medical practitioner who is for the time being appointed in writing by the Executive for the purposes of these Regulations;

(a) S.I. 1991/2289.
(b) 1972 c. 68.
(c) 1974 c. 37; sections 15(1), 43 and 52 were amended by the Employment Protection Act 1975 (c. 71), Schedule 15, paragraphs 6, 12 and 17 respectively; section 51A was added and section 52 was amended by the Police (Health and Safety) Act 1997 (c. 42), sections 1 and 2 respectively.

"approved" means approved for the time being in writing for the purposes of these Regulations by the Health and Safety Commission or the Executive, as the case may be, and published in such form as the Health and Safety Commission or the Executive respectively considers appropriate;

"approved dosimetry service" means, subject to regulation 39(3) (which relates to transitional provisions), a dosimetry service approved in accordance with regulation 35;

"calendar year" means a period of 12 calendar months beginning with the 1st January;

"classified person" means—

(a) a person designated as such pursuant to regulation 20(1); and

(b) in the case of an outside worker employed by an undertaking in Northern Ireland or in another member State, a person who has been designated as a Category A exposed worker within the meaning of Article 21 of the Directive;

"comforter and carer" means an individual who (other than as part of his occupation) knowingly and willingly incurs an exposure to ionising radiation resulting from the support and comfort of another person who is undergoing or who has undergone any medical exposure;

"contamination" means the contamination by any radioactive substance of any surface (including any surface of the body or clothing) or any part of absorbent objects or materials or the contamination of liquids or gases by any radioactive substance;

"controlled area" means—

(a) in the case of an area situated in Great Britain, an area which has been so designated in accordance with regulation 16(1); and

(b) in the case of an area situated in Northern Ireland or in another member State, an area subject to special rules for the purposes of protection against ionising radiation and to which access is controlled as specified in Article 19 of the Directive;

"the Directive" means Council Directive 96/29/Euratom(**a**) laying down basic safety standards for the protection of the health of workers and the general public against the dangers arising from ionising radiation;

"dose" means, in relation to ionising radiation, any dose quantity or sum of dose quantities mentioned in Schedule 4;

"dose assessment" means the dose assessment made and recorded by an approved dosimetry service in accordance with regulation 21;

"dose constraint" means a restriction on the prospective doses to individuals which may result from a defined source;

"dose limit" means, in relation to persons of a specified class, the limit on effective dose or equivalent dose specified in Schedule 4 in relation to a person of that class;

"dose rate" means, in relation to a place, the rate at which a person or part of a person would receive a dose of ionising radiation from external radiation if he were at that place being a dose rate at that place averaged over one minute;

"dose record" means, in relation to a person, the record of the doses received by that person as a result of his exposure to ionising radiation, being the record made and maintained on behalf of the employer by the approved dosimetry service in accordance with regulation 21;

"employment medical adviser" means an employment medical adviser appointed under section 56 of the Health and Safety at Work etc. Act 1974;

"the Executive" means the Health and Safety Executive;

"external radiation" means, in relation to a person, ionising radiation coming from outside the body of that person;

"health record" means, subject to regulation 39(7) (which relates to transitional provisions), in relation to an employee, the record of medical surveillance of that employee maintained by the employer in accordance with regulation 24(3);

"internal radiation" means, in relation to a person, ionising radiation coming from inside the body of that person;

"ionising radiation" means the transfer of energy in the form of particles or electromagnetic waves of a wavelength of 100 nanometres or less or a frequency of 3×10^{15} hertz or more capable of producing ions directly or indirectly;

(**a**) OJ No. L159, 29.6.96, p.1.

"licensee" has the meaning assigned to it by section 26(1) of the Nuclear Installations Act 1965(a);

"local rules" means rules made in accordance with regulation 17;

"maintained", where the reference is to maintaining plant, apparatus, equipment or facilities, means maintained in an efficient state, in efficient working order and good repair;

"medical exposure" means exposure of a person to ionising radiation for the purpose of his medical or dental examination or treatment which is conducted under the direction of a suitably qualified person and includes any such examination for legal purposes and any such examination or treatment conducted for the purposes of research;

"member State" means a member State of the Communities;

"outside worker" means a classified person who carries out services in the controlled area of any employer (other than the controlled area of his own employer);

"overexposure" means any exposure of a person to ionising radiation to the extent that the dose received by that person causes a dose limit relevant to that person to be exceeded or, in relation to regulation 26(2), causes a proportion of a dose limit relevant to any employee to be exceeded;

"practice" means work involving—

(a) the production, processing, handling, use, holding, storage, transport or disposal of radioactive substances; or

(b) the operation of any electrical equipment emitting ionising radiation and containing components operating at a potential difference of more than 5kV,

which can increase the exposure of individuals to radiation from an artificial source, or from a radioactive substance containing naturally occurring radionuclides which are processed for their radioactive, fissile or fertile properties;

"radiation accident" means an accident where immediate action would be required to prevent or reduce the exposure to ionising radiation of employees or any other persons;

"radiation employer" means an employer who in the course of a trade, business or other undertaking carries out work with ionising radiation and, for the purposes of regulations 5, 6 and 7, includes an employer who intends to carry out such work;

"radiation passbook" means—

(a) in the case of an outside worker employed by an employer in Great Britain—

  (i) a passbook approved by the Executive for the purpose of these Regulations; or

  (ii) a passbook to which regulation 39(4) (transitional provisions) applies; and

(b) in the case of an outside worker employed by an employer in Northern Ireland or in another member State, a passbook authorised by the competent authority for Northern Ireland or that member State, as the case may be;

"radiation protection adviser" means, subject to regulation 39(6) (which relates to transitional provisions), an individual who, or a body which, meets such criteria of competence as may from time to time be specified in writing by the Executive;

"radioactive substance" means any substance which contains one or more radionuclides whose activity cannot be disregarded for the purposes of radiation protection;

"sealed source" means a source containing any radioactive substance whose structure is such as to prevent, under normal conditions of use, any dispersion of radioactive substances into the environment, but it does not include any radioactive substance inside a nuclear reactor or any nuclear fuel element;

"short-lived daughters of radon 222" means polonium 218, lead 214, bismuth 214 and polonium 214;

"supervised area" means an area which has been so designated by the employer in accordance with regulation 16(3);

"trainee" means a person aged 16 years or over (including a student) who is undergoing instruction or training which involves operations which would, in the case of an employee, be work with ionising radiation;

(a) 1965 c. 57; relevant amending instruments are S.I. 1974/2056 and S.I. 1990/1918.

"transport" means, in relation to a radioactive substance, carriage of that substance on a road within the meaning of, in relation to England and Wales, section 192 of the Road Traffic Act 1988(**a**) and, in relation to Scotland, the Roads (Scotland) Act 1984(**b**) or through another public place (whether on a conveyance or not), or by rail, inland waterway, sea or air and, in the case of transport on a conveyance, a substance shall be deemed as being transported from the time that it is loaded onto the conveyance for the purpose of transporting it until it is unloaded from that conveyance, but a substance shall not be considered as being transported if—

(a) it is transported by means of a pipeline or similar means; or

(b) it forms an integral part of a conveyance and is used in connection with the operation of that conveyance;

"woman of reproductive capacity" means a woman who is made subject to the additional dose limit for a woman of reproductive capacity specified in paragraphs 5 and 11 of Schedule 4 by an entry in her health record made by an appointed doctor or employment medical adviser;

"work with ionising radiation" means work to which these Regulations apply by virtue of regulation 3(1).

(2) In these Regulations, unless the context otherwise requires, any reference to—

(a) an employer includes a reference to a self-employed person and any duty imposed by these Regulations on an employer in respect of his employee shall extend to a self-employed person in respect of himself;

(b) an employee includes a reference to—

   (i) a self-employed person, and

   (ii) a trainee who but for the operation of this sub-paragraph and paragraph (3) would not be classed as an employee;

(c) exposure to ionising radiation is a reference to exposure to ionising radiation arising from work with ionising radiation;

(d) a person entering, remaining in or working in a controlled or supervised area includes a reference to any part of a person entering, remaining in or working in any such area.

(3) For the purposes of these Regulations and Part I of the Health and Safety at Work etc. Act 1974—

(a) the word "work" shall be extended to include any instruction or training which a person undergoes as a trainee and the meaning of "at work" shall be extended accordingly; and

(b) a trainee shall, while he is undergoing instruction or training in respect of work with ionising radiation, be treated as the employee of the person whose undertaking (whether for profit or not) is providing that instruction or training and that person shall be treated as the employer of that trainee except that the duties to the trainee imposed upon the person providing instruction or training shall only extend to matters under the control of that person.

(4) In these Regulations, where reference is made to a quantity specified in Schedule 8, that quantity shall be treated as being exceeded if—

(a) where only one radionuclide is involved, the quantity of that radionuclide exceeds the quantity specified in the appropriate entry in Schedule 8; or

(b) where more than one radionuclide is involved, the quantity ratio calculated in accordance with Part II of Schedule 8 exceeds one.

(5) Nothing in these Regulations shall be construed as preventing a person from entering or remaining in a controlled area or a supervised area where that person enters or remains in any such area—

(a) in the due exercise of a power of entry conferred on him by or under any enactment; or

(b) for the purpose of undergoing a medical exposure.

(**a**) 1988 c. 52.
(**b**) 1984 c. 54.

(6) In these Regulations—
   (a) any reference to an effective dose means the sum of the effective dose to the whole body from external radiation and the committed effective dose from internal radiation; and
   (b) any reference to equivalent dose to a human tissue or organ includes the committed equivalent dose to that tissue or organ from internal radiation.

(7) In these Regulations—
   (a) a numbered Regulation or Schedule is a reference to the Regulation or Schedule in these Regulations so numbered;
   (b) a numbered paragraph is a reference to the paragraph so numbered in the Regulation or Schedule in which that reference appears.

Application

3.—(1) Subject to the provisions of this regulation and to regulation 6(1), these Regulations shall apply to—
   (a) any practice;
   (b) any work (other than a practice) carried out in an atmosphere containing radon 222 gas at a concentration in air, averaged over any 24 hour period, exceeding 400 Bq m^{-3} except where the concentration of the short-lived daughters of radon 222 in air averaged over any 8 hour working period does not exceed 6.24×10^{-7} Jm^{-3}; and
   (c) any work (other than work referred to in sub-paragraphs (a) and (b) above) with any radioactive substance containing naturally occurring radionuclides.

(2) The following Regulations shall not apply where the only work being undertaken is that referred to in sub-paragraph (b) of paragraph (1), namely regulations 23, 27 to 30, 32 and 33.

(3) The following regulations shall not apply in relation to persons undergoing medical exposures, namely regulations 7, 8, 11, 16 to 18, 23, 25, 31(1) and 34(1).

(4) Regulation 11 shall not apply in relation to any comforter and carer.

(5) In the case of an outside worker (working in a controlled area situated in Great Britain) employed by an employer established in Northern Ireland or in another member State, it shall be sufficient compliance with regulation 21 (dose assessment and recording) and regulation 24 (medical surveillance) if the employer complies with—
   (a) where the employer is established in Northern Ireland, regulations 13 and 16 of the Ionising Radiations Regulations (Northern Ireland) 1985(**a**); or
   (b) where the employer is established in another member State, the legislation in that State implementing Chapters II and III of Title VI of the Directive where such legislation exists.

Duties under the Regulations

4.—(1) Any duty imposed by these Regulations on an employer in respect of the exposure to ionising radiation of persons other than his employees shall be imposed only in so far as the exposure of those persons to ionising radiation arises from work with ionising radiation undertaken by that employer.

(2) Duties under these Regulations imposed upon the employer shall also be imposed upon—
   (a) the manager of a mine (within the meaning of section 180 of the Mines and Quarries Act 1954(**b**)); and
   (b) the operator of a quarry (within the meaning of the Quarries Regulations 1999(**c**)),
in so far as those duties relate to the mine or part of the mine of which he is the manager or the quarry of which he is the operator and to matters within his control.

(**a**) S.R. 1985/273.
(**b**) 1954 c. 70; section 180 was amended by S.I. 1993/1897.
(**c**) S.I. 1999/2024.

(3) Subject to regulation 6(1)(b), duties under these Regulations imposed upon the employer shall also be imposed on the holder of a nuclear site licence under the Nuclear Installations Act 1965(**a**) in so far as those duties relate to the licensed site.

PART II

GENERAL PRINCIPLES AND PROCEDURES

Authorisation of specified practices

5.—(1) Subject to paragraph (2), a radiation employer shall not, except in accordance with a prior authorisation granted by the Executive in writing for the purposes of this paragraph, carry out the following practices—

- (a) the use of electrical equipment intended to produce x-rays for the purpose of—
    - (i) industrial radiography;
    - (ii) the processing of products;
    - (iii) research; or
    - (iv) the exposure of persons for medical treatment; or
- (b) the use of accelerators, except electron microscopes.

(2) Paragraph (1) shall not apply in respect of any practice of a type which is for the time being authorised by the Executive where such practice is or is to be carried out in accordance with such conditions as may from time to time be approved by the Executive in respect of that type of practice.

(3) An authorisation granted under paragraph (1) may be granted subject to conditions and with or without limit of time and may be revoked in writing at any time.

(4) Where an authorisation has been granted pursuant to paragraph (1) and the radiation employer to whom the authorisation was granted subsequently makes a material change to the circumstances relating to that authorisation, that change shall forthwith be notified to the Executive by the radiation employer.

(5) A radiation employer to whom this regulation applies and who is aggrieved by—

- (a) a decision of the Executive—
    - (i) refusing to grant an authorisation under paragraph (1);
    - (ii) imposing a limit of time upon an authorisation granted under paragraph (1); or
    - (iii) revoking an authorisation under paragraph (3); or
- (b) the terms of any conditions attached to the authorisation by the Executive under paragraph (3),

may appeal to the Secretary of State.

(6) Sub-sections (2) to (6) of section 44 of the 1974 Act shall apply for the purposes of paragraph (5) as they apply to an appeal under section 44(1) of that Act.

(7) The Health and Safety Licensing Appeals (Hearings Procedure) Rules 1974(**b**), as respects England and Wales, and the Health and Safety Licensing Appeals (Hearings Procedure) (Scotland) Rules 1974(**c**), as respects Scotland, shall apply to an appeal under paragraph (5) as they apply to an appeal under sub-section (1) of the said section 44, but with the modification that references to a licensing authority are to be read as references to the Executive.

Notification of specified work

6.—(1) This regulation shall apply to work with ionising radiation except—

- (a) work specified in Schedule 1; and
- (b) work carried on at a site licensed under section 1 of the Nuclear Installations Act 1965.

(**a**) 1965 c. 57; relevant amending instruments are S.I. 1974/2056 and S.I. 1990/1918.
(**b**) S.I. 1974/2040.
(**c**) S.I. 1974/2068.

(2) Subject to paragraphs (7) and (8) and to regulation 39(1) (which relates to transitional provisions), a radiation employer shall not for the first time carry out work with ionising radiation to which this regulation applies unless at least 28 days before commencing that work or before such shorter time as the Executive may agree he has notified the Executive of his intention to carry out that work and has provided the Executive with the particulars specified in Schedule 2.

(3) Where a radiation employer has notified work in accordance with paragraph (2), the Executive may, by notice in writing served on him, require that radiation employer to provide such additional particulars of that work as it may reasonably require, being any or all of the particulars specified in Schedule 3, and in such a case the radiation employer shall provide those particulars by such time as is specified in the notice or by such other time as the Executive may subsequently agree.

(4) A notice under paragraph (3) may require the radiation employer to notify the Executive of any of the particulars specified in Schedule 3 before each occasion on which he commences work with ionising radiation.

(5) Where a radiation employer has notified work in accordance with paragraph (2) and subsequently makes a material change in that work which would affect the particulars so notified, he shall forthwith notify the Executive of that change.

(6) Nothing in paragraph (5) shall be taken as requiring the cessation of the work to be notified in accordance with that paragraph except where the site or any part of the site in which the work was carried on has been or is to be vacated.

(7) Where the only work being undertaken is work referred to in regulation 3(1)(b) or (c), it shall be a sufficient compliance with paragraph (2) if the radiation employer having control of the premises where the work is carried on makes the notification required by that paragraph forthwith after the work has commenced.

(8) In relation to work involving the care of a person to whom a radioactive medicinal product (within the meaning of the Medicines (Administration of Radioactive Substances) Regulations 1978(a)) has been administered, it shall be sufficient compliance with paragraph (2) if the notification required by that paragraph is given as soon as is practicable before the carrying out of that work.

(9) Where in respect of work with ionising radiation carried out prior to the coming into force of these Regulations notification has been given to the Executive pursuant to any statutory requirement, the provisions of this regulation shall apply to such notification as if that notification had been given in accordance with paragraph (2).

Prior risk assessment etc.

7.—(1) Before a radiation employer commences a new activity involving work with ionising radiation in respect of which no risk assessment has been made by him, he shall make a suitable and sufficient assessment of the risk to any employee and other person for the purpose of identifying the measures he needs to take to restrict the exposure of that employee or other person to ionising radiation.

(2) Without prejudice to paragraph (1), a radiation employer shall not carry out work with ionising radiation unless he has made an assessment sufficient to demonstrate that—
  (a) all hazards with the potential to cause a radiation accident have been identified; and
  (b) the nature and magnitude of the risks to employees and other persons arising from those hazards have been evaluated.

(3) Where the assessment made for the purposes of this regulation shows that a radiation risk to employees or other persons exists from an identifiable radiation accident, the radiation employer shall take all reasonably practicable steps to—
  (a) prevent any such accident;
  (b) limit the consequences of any such accident which does occur; and
  (c) provide employees with the information, instruction and training, and with the equipment necessary, to restrict their exposure to ionising radiation.

(a) S.I. 1978/1006.

(4) The requirements of this regulation are without prejudice to the requirements of regulation 3 (Risk assessment) of the Management of Health and Safety at Work Regulations 1992(**a**).

Restriction of exposure

8.—(1) Every radiation employer shall, in relation to any work with ionising radiation that he undertakes, take all necessary steps to restrict so far as is reasonably practicable the extent to which his employees and other persons are exposed to ionising radiation.

(2) Without prejudice to the generality of paragraph (1), a radiation employer shall—

(a) so far as is reasonably practicable achieve the restriction of exposure to ionising radiation required under that paragraph by means of engineering controls and design features and in addition by the provision and use of safety features and warning devices; and

(b) in addition to sub-paragraph (a) above, provide such systems of work as will, so far as is reasonably practicable, restrict the exposure to ionising radiation of employees and other persons; and

(c) in addition to sub-paragraphs (a) and (b) above, where it is reasonably practicable to further restrict exposure to ionising radiation by means of personal protective equipment, provide employees or other persons with adequate and suitable personal protective equipment (including respiratory protective equipment) unless the use of personal protective equipment of a particular kind is not appropriate having regard to the nature of the work or the circumstances of the particular case.

(3) Where it is appropriate to do so at the planning stage of radiation protection, dose constraints shall be used in restricting exposure to ionising radiation pursuant to paragraph (1).

(4) An employer who provides any system of work or personal protective equipment pursuant to this regulation shall take all reasonable steps to ensure that it is properly used or applied as the case may be.

(5) Without prejudice to paragraph (1), a radiation employer shall ensure, that—

(a) in relation to an employee who is pregnant, the conditions of exposure are such that, after her employer has been notified of the pregnancy, the equivalent dose to the foetus is unlikely to exceed 1mSv during the remainder of the pregnancy; and

(b) in relation to an employee who is breastfeeding, the conditions of exposure are restricted so as to prevent significant bodily contamination of that employee.

(6) Nothing in paragraph (5) shall require the radiation employer to take any action in relation to an employee until she has notified her employer in writing that she is pregnant or breastfeeding and the radiation employer has been made aware, or should reasonably have been expected to be aware, of that fact.

(7) Every employer shall, for the purpose of determining whether the requirements of paragraph (1) are being met, ensure that an investigation is carried out forthwith when the effective dose of ionising radiation received by any of his employees for the first time in any calendar year exceeds 15mSv or such other lower effective dose as the employer may specify, which dose shall be specified in writing in local rules made pursuant to regulation 17(1) or, where local rules are not required, by other suitable means.

Personal protective equipment

9.—(1) Any personal protective equipment provided by an employer pursuant to regulation 8 shall comply with any provision in the Personal Protective Equipment (EC Directive) Regulations 1992(**b**) which is applicable to that item of personal protective equipment.

(**a**) S.I. 1992/2051.
(**b**) S.I. 1992/3139; as amended by S.I. 1993/3074, S.I. 1994/2326 and S.I. 1996/3039.

(2) Where in the case of respiratory protective equipment no provision of the Regulations referred to in paragraph (1) applies, that respiratory protective equipment shall satisfy the requirements of regulation 8 only if it is of a type, or conforms to a standard, approved in either case by the Executive.

(3) Every radiation employer shall ensure that appropriate accommodation is provided for personal protective equipment when it is not being worn.

Maintenance and examination of engineering controls etc. and personal protective equipment

10.—(1) A radiation employer who provides any engineering control, design feature, safety feature or warning device to meet the requirements of regulation 8(2)(a) shall ensure—

(a) that any such control, feature or device is properly maintained; and

(b) where appropriate, that thorough examinations and tests of such controls, features or devices are carried out at suitable intervals.

(2) Every radiation employer shall ensure that all personal protective equipment provided pursuant to regulation 8 is, where appropriate, thoroughly examined at suitable intervals and is properly maintained and that, in the case of respiratory protective equipment, a suitable record of that examination is made and kept for at least two years from the date on which the examination was made and that the record includes a statement of the condition of the equipment at the time of the examination.

Dose limitation

11.—(1) Subject to paragraph (2) and to paragraph 5 of Schedule 4, every employer shall ensure that his employees and other persons within a class specified in Schedule 4 are not exposed to ionising radiation to an extent that any dose limit specified in Part I of that Schedule for such class of person is exceeded in any calendar year.

(2) Where an employer is able to demonstrate in respect of any employee that the dose limit specified in paragraph 1 of Part I of Schedule 4 is impracticable having regard to the nature of the work undertaken by that employee, the employer may in respect of that employee apply the dose limits set out in paragraphs 9 to 11 of that Schedule and in such case the provisions of Part II of the Schedule shall have effect.

Contingency plans

12.—(1) Where an assessment made in accordance with regulation 7 shows that a radiation accident is reasonably foreseeable (having regard to the steps taken by the radiation employer under paragraph (3) of that regulation), the radiation employer shall prepare a contingency plan designed to secure, so far as is reasonably practicable, the restriction of exposure to ionising radiation and the health and safety of persons who may be affected by such accident.

(2) The radiation employer shall ensure that—

(a) where local rules are required for the purposes of regulation 17, a copy of the contingency plan made in pursuance of paragraph (1) is identified in those rules and incorporated into them by way of summary or reference;

(b) any employee under his control who may be involved with or may be affected by arrangements in the plan has been given suitable and sufficient instructions and where appropriate issued with suitable dosemeters or other devices obtained in either case from the approved dosimetry service with which the radiation employer has entered into an arrangement under regulation 21; and

(c) where appropriate, rehearsals of the arrangements in the plan are carried out at suitable intervals.

PART III

ARRANGEMENTS FOR THE MANAGEMENT OF RADIATION PROTECTION

Radiation protection adviser

13.—(1) Subject to paragraph (3), every radiation employer shall consult such suitable radiation protection advisers as are necessary for the purpose of advising the radiation employer as to the observance of these Regulations and shall, in any event, consult one or more suitable radiation protection advisers with regard to those matters which are set out in Schedule 5.

(2) Where a radiation protection adviser is consulted pursuant to the requirements of paragraph (1) (other than in respect of the observance of that paragraph), the radiation employer shall appoint that radiation protection adviser in writing and shall include in that appointment the scope of the advice which the radiation protection adviser is required to give.

(3) Nothing in paragraph (1) shall require a radiation employer to consult a radiation protection adviser where the only work with ionising radiation undertaken by that employer is work specified in Schedule 1.

(4) The radiation employer shall provide any radiation protection adviser appointed by him with adequate information and facilities for the performance of his functions.

Information, instruction and training

14. Every employer shall ensure that—
   (a) those of his employees who are engaged in work with ionising radiation are given appropriate training in the field of radiation protection and receive such information and instruction as is suitable and sufficient for them to know—
      (i) the risks to health created by exposure to ionising radiation;
      (ii) the precautions which should be taken; and
      (iii) the importance of complying with the medical, technical and administrative requirements of these Regulations;
   (b) adequate information is given to other persons who are directly concerned with the work with ionising radiation carried on by the employer to ensure their health and safety so far as is reasonably practicable; and
   (c) those female employees of that employer who are engaged in work with ionising radiation are informed of the possible risk arising from ionising radiation to the foetus and to a nursing infant and of the importance of those employees informing the employer in writing as soon as possible—
      (i) after becoming aware of their pregnancy; or
      (ii) if they are breast feeding.

Co-operation between employers

15. Where work with ionising radiation undertaken by one employer is likely to give rise to the exposure to ionising radiation of the employee of another employer, the employers concerned shall co-operate by the exchange of information or otherwise to the extent necessary to ensure that each such employer is enabled to comply with the requirements of these Regulations in so far as his ability to comply depends upon such co-operation.

PART IV
DESIGNATED AREAS

Designation of controlled or supervised areas

16.—(1) Every employer shall designate as a controlled area any area under his control which has been identified by an assessment made by him (whether pursuant to regulation 7 or otherwise) as an area in which—

(a) it is necessary for any person who enters or works in the area to follow special procedures designed to restrict significant exposure to ionising radiation in that area or prevent or limit the probability and magnitude of radiation accidents or their effects; or

(b) any person working in the area is likely to receive an effective dose greater than 6mSv a year or an equivalent dose greater than three-tenths of any relevant dose limit referred to in Schedule 4 in respect of an employee aged 18 years or above.

(2) An employer shall not intentionally create in any area conditions which would require that area to be designated as a controlled area unless that area is for the time being under the control of that employer.

(3) An employer shall designate as a supervised area any area under his control, not being an area designated as a controlled area—

(a) where it is necessary to keep the conditions of the area under review to determine whether the area should be designated as a controlled area; or

(b) in which any person is likely to receive an effective dose greater than 1mSv a year or an equivalent dose greater than one-tenth of any relevant dose limit referred to in Schedule 4 in respect of an employee aged 18 years or above.

Local rules and radiation protection supervisors

17.—(1) For the purposes of enabling work with ionising radiation to be carried on in accordance with the requirements of these Regulations, every radiation employer shall, in respect of any controlled area or, where appropriate having regard to the nature of the work carried out there, any supervised area, make and set down in writing such local rules as are appropriate to the radiation risk and the nature of the operations undertaken in that area.

(2) The radiation employer shall take all reasonable steps to ensure that any local rules made pursuant to paragraph (1) and which are relevant to the work being carried out are observed.

(3) The radiation employer shall ensure that such of those rules made pursuant to paragraph (1) as are relevant are brought to the attention of those employees and other persons who may be affected by them.

(4) The radiation employer shall—

(a) appoint one or more suitable radiation protection supervisors for the purpose of securing compliance with these Regulations in respect of work carried out in any area made subject to local rules pursuant to paragraph (1); and

(b) set down in the local rules the names of such individuals so appointed.

Additional requirements for designated areas

18.—(1) Every employer who designates any area as a controlled or supervised area shall ensure that any such designated area is adequately described in local rules and that—

(a) in the case of any controlled area—

(i) the area is physically demarcated or, where this is not reasonably practicable, delineated by some other suitable means; and

(ii) suitable and sufficient signs are displayed in suitable positions indicating that the area is a controlled area, the nature of the radiation sources in that area and the risks arising from such sources; and

(b) in the case of any supervised area, suitable and sufficient signs giving warning of the supervised area are displayed, where appropriate, in suitable positions indicating the nature of the radiation sources and the risks arising from such sources.

(2) The employer who has designated an area as a controlled area shall not permit any employee or other person to enter or remain in such an area unless that employee or other person—

   (a) being a person other than an outside worker, is a classified person;

   (b) being an outside worker, is a classified person in respect of whom the employer has taken all reasonable steps to ensure that the person—

      (i) is subject to individual dose assessment pursuant to regulation 21;

      (ii) has been provided with and has been trained to use any personal protective equipment that may be necessary pursuant to regulation 8(2)(c);

      (iii) has received any specific training required pursuant to regulation 14; and

      (iv) has been certified fit for the work with ionising radiation which he is to carry out pursuant to regulation 24; or

   (c) not being a classified person, enters or remains in the area in accordance with suitable written arrangements for the purpose of ensuring that—

      (i) in the case of an employee aged 18 years or over, he does not receive in any calendar year a cumulative dose of ionising radiation which would require that employee to be designated as a classified person; or

      (ii) in the case of any other person, he does not receive in any calendar year a dose of ionising radiation exceeding any relevant dose limit.

(3) An employer who has designated an area as a controlled area shall not permit a person to enter or remain in such area in accordance with the written arrangements under paragraph (2)(c), unless he can demonstrate, by personal dose monitoring or other suitable measurements, that the doses are restricted in accordance with that sub-paragraph.

(4) An employer who has designated an area as a controlled area shall, in relation to an outside worker, ensure that—

   (a) the outside worker is subject to arrangements for estimating the dose of ionising radiation he receives whilst in the controlled area;

   (b) as soon as is reasonably practicable after the services carried out by that outside worker in that controlled area are completed, an estimate of the dose received by that worker is entered into his radiation passbook; and

   (c) when the radiation passbook of the outside worker is in the possession of that employer, the passbook is made available to that worker upon request.

(5) The employer who carries out the monitoring or measurements pursuant to paragraph (3) shall keep the results of the monitoring or measurements referred to in that paragraph for a period of two years from the date they were recorded and shall, at the request of the person to whom the monitoring or measurements relate and on reasonable notice being given make the results available to that person.

(6) In any case where there is a significant risk of the spread of radioactive contamination from a controlled area, the employer who has designated that area as a controlled area shall make adequate arrangements to restrict, so far as is reasonably practicable, the spread of such contamination.

(7) Without prejudice to the generality of paragraph (6), the arrangements required by that paragraph shall, where appropriate, include—

   (a) the provision of suitable and sufficient washing and changing facilities for persons who enter or leave any controlled or supervised area;

   (b) the proper maintenance of such washing and changing facilities;

   (c) the prohibition of eating, drinking or smoking or similar activity likely to result in the ingestion of a radioactive substance by any employee in a controlled area; and

   (d) the means for monitoring for contamination any person, article or goods leaving a controlled area.

Monitoring of designated areas

19.—(1) Every employer who designates an area as a controlled or supervised area shall take such steps as are necessary (otherwise than by use of assessed doses of individuals), having regard to the nature and extent of the risks resulting from exposure to ionising radiation, to ensure that levels of ionising radiation are adequately monitored for each such area and that working conditions in those areas are kept under review.

(2) The employer upon whom a duty is imposed by paragraph (1) shall provide suitable and sufficient equipment for carrying out the monitoring required by that paragraph, which equipment shall—

   (a) be properly maintained so that it remains fit for the purpose for which it was intended; and
   (b) be adequately tested and examined at appropriate intervals.

(3) Equipment provided pursuant to paragraph (2) shall not be or remain suitable unless—

   (a) the performance of the equipment has been established by adequate tests before it has first been used; and
   (b) the tests and examinations carried out pursuant to paragraph (2) and sub-paragraph (a) above have been carried out by or under the supervision of a qualified person.

(4) The employer upon whom a duty is imposed by paragraph (1) shall—

   (a) make suitable records of the results of the monitoring carried out in accordance with paragraph (1) and of the tests carried out in accordance with paragraphs (2) and (3);
   (b) ensure that the records of the tests carried out pursuant to sub-paragraph (a) above are authorised by a qualified person; and
   (c) keep the records referred to in sub-paragraph (a) above, or copies thereof, for at least 2 years from the respective dates on which they were made.

PART V

CLASSIFICATION AND MONITORING OF PERSONS

Designation of classified persons

20.—(1) Subject to paragraph (2), the employer shall designate as classified persons those of his employees who are likely to receive an effective dose in excess of 6mSv per year or an equivalent dose which exceeds three-tenths of any relevant dose limit and shall forthwith inform those employees that they have been so designated.

(2) The employer shall not designate an employee as a classified person unless—

   (a) that employee is aged 18 years or over; and
   (b) an appointed doctor or employment medical adviser has certified in the health record that that employee is fit for the work with ionising radiation which he is to carry out.

(3) The employer may cease to treat an employee as a classified person only at the end of a calendar year except where—

   (a) an appointed doctor or employment medical adviser so requires; or
   (b) the employee is no longer employed by the same employer in a capacity which is likely to result in significant exposure to ionising radiation during the remainder of the relevant calendar year.

Dose assessment and recording

21.—(1) Every employer shall ensure that—

   (a) in respect of each of his employees who is designated as a classified person, an assessment is made of all doses of ionising radiation received by such employee which are likely to be significant; and
   (b) such assessments are recorded.

(2) For the purposes of paragraph (1), the employer shall make suitable arrangements with one or more approved dosimetry service for—

    (a) the making of systematic assessments of such doses by the use of suitable individual measurement for appropriate periods or, where individual measurement is inappropriate, by means of other suitable measurements; and

    (b) the making and maintenance of dose records relating to each classified person.

(3) For the purposes of paragraph (2)(b), the arrangements that the employer makes with the approved dosimetry service shall include requirements for that service—

    (a) to keep the records made and maintained pursuant to the arrangements or a copy thereof until the person to whom the record relates has or would have attained the age of 75 years but in any event for at least 50 years from when they were made;

    (b) to provide the employer at appropriate intervals with suitable summaries of the dose records maintained in accordance with sub-paragraph (a) above;

    (c) when required by the employer, to provide him with such copies of the dose record relating to any of his employees as the employer may require;

    (d) when required by the employer, to make a record of the information concerning the dose assessment relating to a classified person who ceases to be an employee of the employer, and to send that record to the Executive and a copy thereof to the employer forthwith, and a record so made is referred to in this regulation as a "termination record";

    (e) within 3 months, or such longer period as the Executive may agree, of the end of each calendar year to send to the Executive summaries of all current dose records relating to that year;

    (f) when required by the Executive, to provide it with copies of any dose records;

    (g) where a dose is estimated pursuant to regulation 22, to make an entry in a dose record and retain the summary of the information used to estimate that dose;

    (h) where the employer employs an outside worker, to provide, where appropriate, a current radiation passbook in respect of that outside worker; and

    (i) where the employer employs an outside worker who works in Northern Ireland or another member State, maintain a continuing record of the assessment of the dose received by that outside worker when working in such place.

(4) The employer shall provide the approved dosimetry service with such information concerning his employees as is necessary for the approved dosimetry service to comply with the arrangements made for the purposes of paragraph (2).

(5) An employer shall—

    (a) ensure that each outside worker employed by him is provided with a current individual radiation passbook which shall not be transferable to any other worker and in which shall be entered the particulars set out in Schedule 6; and

    (b) make suitable arrangements to ensure that the particulars entered in the radiation passbook are kept up-to-date during the continuance of the employment of the outside worker by that employer.

(6) The employer shall—

    (a) at the request of a classified person employed by him (or of a person formerly employed by him as a classified person) and on reasonable notice being given, obtain (where necessary) from the approved dosimetry service and make available to that person—

        (i) a copy of the dose summary provided for the purpose of paragraph (3)(b) relating to that person and made within a period of 2 years preceding the request; and

        (ii) a copy of the dose record of that person; and

    (b) when a classified person ceases to be employed by the employer, take all reasonable steps to provide to that person a copy of his termination record.

(7) The employer shall keep a copy of the summary of the dose record received from the approved dosimetry service for at least 2 years from the end of the calendar year to which the summary relates.

Estimated doses and special entries

22.—(1) Where a dosemeter or other device is used to make any individual measurement under regulation 21(2) and that dosemeter or device is lost, damaged or destroyed or it is not practicable to assess the dose received by a classified person over any period, the employer shall make an adequate investigation of the circumstances of the case with a view to estimating the dose received by that person during that period and either—

  (a) in a case where there is adequate information to estimate the dose received by that person, shall send to the approved dosimetry service an adequate summary of the information used to estimate that dose and shall arrange for the approved dosimetry service to enter the estimated dose in the dose record of that person; or

  (b) in a case where there is inadequate information to estimate the dose received by the classified person, shall arrange for the approved dosimetry service to enter a notional dose in the dose record of that person which shall be the proportion of the total annual dose limit for the relevant period,

and in either case the employer shall take reasonable steps to inform the classified person of that entry and arrange for the approved dosimetry service to identify the entry in the dose record as an estimated dose or a notional dose as the case may be.

(2) The employer shall, at the request of the classified person (or a person formerly employed by that employer as a classified person) to whom the investigation made under paragraph (1) relates and on reasonable notice being given, make available to that person a copy of the summary sent to the approved dosimetry service under sub-paragraph (a) of paragraph (1).

(3) Subject to paragraphs (5) and (8), where an employer has reasonable cause to believe that the dose received by a classified person is much greater or much less than that shown in the relevant entry of the dose record, he shall make an adequate investigation of the circumstances of the exposure of that person to ionising radiation and, if that investigation confirms his belief, the employer shall, where there is adequate information to estimate the dose received by the employee—

  (a) send to the approved dosimetry service an adequate summary of the information used to estimate that dose;

  (b) arrange for the approved dosimetry service to enter that estimated dose in the dose record of that person and for the approved dosimetry service to identify the estimated dose in the dose record as a special entry; and

  (c) notify the classified person accordingly.

(4) The employer shall make a report of any investigation carried out under paragraph (3) and shall preserve a copy of that report for a period of 2 years from the date it was made.

(5) Paragraph (3) shall not apply—

  (a) in respect of a classified person subject only to an annual dose limit, more than 12 months after the original entry was made in the record; and

  (b) in any other case, more than 5 years after the original entry was made in the record.

(6) Where a classified person is aggrieved by a decision to replace a recorded dose by an estimated dose pursuant to paragraph (3) he may, by an application in writing to the Executive made within 3 months of the date on which he was notified of the decision, apply for that decision to be reviewed.

(7) Where the Executive concludes (whether as a result of a review carried out pursuant to paragraph (6) or otherwise) that—

  (a) there is reasonable cause to believe the investigation carried out pursuant to paragraph (3) was inadequate; or

  (b) a reasonable estimated dose has not been established,

the employer shall, if so directed by the Executive, re-instate the original entry in the dose record.

(8) The employer shall not, without the consent of the Executive, require the approved dosimetry service to enter an estimated dose in the dose record in any case where—

  (a) the cumulative recorded effective dose is 20mSv or more in one calendar year; or

  (b) the cumulative recorded equivalent dose for the calendar year exceeds a relevant dose limit.

Dosimetry for accidents etc.

23.—(1) Where any accident or other occurrence takes place which is likely to result in a person receiving an effective dose of ionising radiation exceeding 6mSv or an equivalent dose greater than three-tenths of any relevant dose limit, the employer shall—
- (a) in the case of a classified person, arrange for a dose assessment to be made by the approved dosimetry service forthwith;
- (b) in the case of an employee to whom a dosemeter or other device has been issued in accordance with regulation 12(2), arrange for that dosemeter or device to be examined and for the dose received to be assessed by the approved dosimetry service as soon as possible;
- (c) in any other case, arrange for the dose to be assessed by an appropriate means as soon as possible, having regard to the advice of the radiation protection adviser.

(2) In such a case, the employer shall—
- (a) take all reasonably practicable steps to inform each person for whom a dose assessment has been made of the result of that assessment; and
- (b) keep a record of the assessment or a copy thereof until the person to whom the record relates has or would have attained the age of 75 years but in any event for at least 50 years from the date of the relevant accident.

Medical surveillance

24.—(1) This regulation shall apply in relation to—
- (a) classified persons and persons whom an employer intends to designate as classified persons;
- (b) employees who have received an overexposure and are not classified persons;
- (c) employees who are engaged in work with ionising radiation subject to conditions imposed by an appointed doctor or employment medical adviser under paragraph (6).

(2) The employer shall ensure that each of his employees to whom this regulation relates is under adequate medical surveillance by an appointed doctor or employment medical adviser for the purpose of determining the fitness of each employee for the work with ionising radiation which he is to carry out.

(3) The employer shall ensure that a health record, containing the particulars referred to in Schedule 7, in respect of each of his employees to whom this regulation relates is made and maintained and that that record or a copy thereof is kept until the person to whom the record relates has or would have attained the age of 75 years but in any event for at least 50 years from the date of the last entry made in it.

(4) Subject to paragraph (5), the employer shall ensure that there is a valid entry in the health record of each of his employees to whom this regulation relates (other than employees who have received an overexposure and who are not classified persons) made by an appointed doctor or employment medical adviser and an entry in the health record shall be valid—
- (a) for 12 months from the date it was made or treated as made by virtue of paragraph (5);
- (b) for such shorter period as is specified in the entry by the appointed doctor or employment medical adviser; or
- (c) until cancelled by an appointed doctor or employment medical adviser by a further entry in the record.

(5) For the purposes of paragraph (4)(a), a further entry in the health record of the same employee shall, where made not less than 11 months nor more than 13 months after the start of the current period of validity, be treated as if made at the end of that period.

(6) Where the appointed doctor or employment medical adviser has certified in the health record of an employee to whom this regulation relates that in his professional opinion that employee should not be engaged in work with ionising radiation or that he should only be so engaged under conditions he has specified in the health record, the employer shall not permit that employee to be engaged in the work with ionising radiation except in accordance with the conditions, if any, so specified.

(7) Where, for the purpose of carrying out his functions under these Regulations, an appointed doctor or employment medical adviser requires to inspect any workplace, the employer shall permit him to do so.

(8) The employer shall make available to the appointed doctor or employment medical adviser the summary of the dose record kept by the employer pursuant to regulation 21(7) and such other records kept for the purposes of these Regulations as the appointed doctor or employment medical adviser may reasonably require.

(9) Where an employee is aggrieved by a decision recorded in the health record by an appointed doctor or employment medical adviser he may, by an application in writing to the Executive made within 3 months of the date on which he was notified of the decision, apply for that decision to be reviewed in accordance with a procedure approved for the purposes of this paragraph by the Health and Safety Commission, and the result of that review shall be notified to the employee and entered in his health record in accordance with the approved procedure.

Investigation and notification of overexposure

25.—(1) Where a radiation employer suspects or has been informed that any person is likely to have received an overexposure as a result of work carried out by that employer, that employer shall make an immediate investigation to determine whether there are circumstances which show beyond reasonable doubt that no overexposure could have occurred and, unless this is shown, the radiation employer shall—

(a) as soon as practicable notify the suspected overexposure to—
  (i) the Executive;
  (ii) in the case of an employee of some other employer, that other employer; and
  (iii) in the case of his own employee, the appointed doctor or employment medical adviser;
(b) as soon as practicable take reasonable steps to notify the suspected overexposure to the person affected; and
(c) make or arrange for such investigation of the circumstances of the exposure and an assessment of any relevant dose received as is necessary to determine, so far as is reasonably practicable, the measures, if any, required to be taken to prevent a recurrence of such overexposure and shall forthwith notify the results of that investigation and assessment to the persons and authorities mentioned in sub-paragraph (a) above and shall—
  (i) in the case of his employee, forthwith notify that employee of the results of the investigation and assessment, or
  (ii) in the case of a person who is not his employee, where the investigation has shown that that person has received an overexposure, take all reasonable steps to notify him of his overexposure.

(2) A radiation employer who makes any investigation pursuant to paragraph (1) shall make a report of that investigation and shall—

(a) in respect of an immediate investigation, keep that report or a copy thereof for at least 2 years from the date on which it was made; and
(b) in respect of an investigation made pursuant to sub-paragraph (c) of paragraph (1), keep that report or a copy thereof until the person to whom the record relates has or would have attained the age of 75 years but in any event for at least 50 years from the date on which it was made.

(3) Where the person who received the overexposure is an employee who has a dose record, his employer shall arrange for the assessment of the dose received to be entered into that dose record.

Dose limitation for overexposed employees

26.—(1) Without prejudice to other requirements of these Regulations and in particular regulation 24(6), where an employee has been subjected to an overexposure paragraph (2) shall apply in relation to the employment of that employee on work with ionising radiation during the remainder of the dose limitation period commencing at the end of the personal dose assessment period in which he was subjected to the overexposure.

(2) The employer shall ensure that an employee to whom this regulation relates does not, during the remainder of the dose limitation period, receive a dose of ionising radiation greater than that proportion of any dose limit which is equal to the proportion that the remaining part of the dose limitation period bears to the whole of that period.

(3) The employer shall inform an employee who has been subjected to an overexposure of the dose limit which is applicable to that employee for the remainder of the relevant dose limitation period.

(4) In this regulation, "dose limitation period" means, as appropriate, a calendar year or the period of five consecutive calendar years.

PART VI

ARRANGEMENTS FOR THE CONTROL OF RADIOACTIVE SUBSTANCES, ARTICLES AND EQUIPMENT

Sealed sources and articles containing or embodying radioactive substances

27.—(1) Where a radioactive substance is used as a source of ionising radiation in work with ionising radiation, the radiation employer shall ensure that, whenever reasonably practicable, the substance is in the form of a sealed source.

(2) The radiation employer shall ensure that the design, construction and maintenance of any article containing or embodying a radioactive substance, including its bonding, immediate container or other mechanical protection, is such as to prevent the leakage of any radioactive substance—

(a) in the case of a sealed source, so far as is practicable; or

(b) in the case of any other article, so far as is reasonably practicable.

(3) Where appropriate, the radiation employer shall ensure that suitable tests are carried out at suitable intervals to detect leakage of radioactive substances from any article to which paragraph (2) applies and the employer shall make a suitable record of each such test and shall retain that record for at least 2 years after the article is disposed of or until a further record is made following a subsequent test to that article.

Accounting for radioactive substances

28. For the purpose of controlling radioactive substances which are involved in work with ionising radiation which he undertakes, every radiation employer shall take such steps as are appropriate to account for and keep records of the quantity and location of those substances and shall keep those records or a copy thereof for at least 2 years from the date on which they were made and, in addition, for at least 2 years from the date of disposal of that radioactive substance.

Keeping and moving of radioactive substances

29.—(1) Every radiation employer shall ensure, so far as is reasonably practicable, that any radioactive substance under his control which is not for the time being in use or being moved, transported or disposed of—

(a) is kept in a suitable receptacle; and

(b) is kept in a suitable store.

(2) Every employer who causes or permits a radioactive substance to be moved (otherwise than by transporting it) shall ensure that, so far as is reasonably practicable, the substance is kept in a suitable receptacle, suitably labelled, while it is being moved.

(3) Nothing in paragraphs (1) or (2) shall apply in relation to a radioactive substance while it is in or on the live body or corpse of a human being.

Notification of certain occurrences

30.—(1) Every radiation employer shall forthwith notify the Executive in any case where a quantity of a radioactive substance which was under his control and which exceeds the quantity specified for that substance in column 4 of Schedule 8—

(a) has been released or is likely to have been released into the atmosphere as a gas, aerosol or dust; or

(b) has been spilled or otherwise released in such a manner as to give rise to significant contamination.

(2) Paragraph (1) shall not apply where such release—
   (a) was in accordance with a registration under section 10 of the Radioactive Substances Act 1993(**a**) or which was exempt from such registration by virtue of section 11 of that Act; or
   (b) was in a manner specified in an authorisation to dispose of radioactive waste under section 13 of the said Act or which was exempt from such authorisation by virtue of section 15 of that Act.

(3) Where a radiation employer has reasonable cause to believe that a quantity of a radioactive substance which exceeds the quantity for that substance specified in column 5 of Schedule 8 and which was under his control is lost or has been stolen, the employer shall forthwith notify the Executive of that loss or theft, as the case may be.

(4) Where a radiation employer suspects or has been informed that an occurrence notifiable under paragraph (1) or (3) may have occurred, he shall make an immediate investigation and, unless that investigation shows that no such occurrence has occurred, he shall forthwith make a notification in accordance with the relevant paragraph.

(5) A radiation employer who makes any investigation in accordance with paragraph (4) shall make a report of that investigation and shall, unless the investigation showed that no such occurrence occurred, keep that report or a copy thereof for at least 50 years from the date on which it was made or, in any other case, for at least 2 years from the date on which it was made.

Duties of manufacturers etc. of articles for use in work with ionising radiation

31.—(1) In the case of articles for use at work, where that work is work with ionising radiation, section 6(1) of the Health and Safety at Work etc. Act 1974(**b**) (which imposes general duties on manufacturers etc. as regards articles and substances for use at work) shall be modified so that any duty imposed on any person by that subsection shall include a duty to ensure that any such article is so designed and constructed as to restrict so far as is reasonably practicable the extent to which employees and other persons are or are likely to be exposed to ionising radiation.

(2) Where a person erects or installs an article for use at work, being work with ionising radiation, he shall—
   (a) where appropriate, undertake a critical examination of the way in which the article was erected or installed for the purpose of ensuring, in particular, that—
      (i) the safety features and warning devices operate correctly; and
      (ii) there is sufficient protection for persons from exposure to ionising radiation;
   (b) consult with the radiation protection adviser appointed by himself or by the radiation employer with regard to the nature and extent of any critical examination and the results of that examination; and
   (c) provide the radiation employer with adequate information about proper use, testing and maintenance of the article.

Equipment used for medical exposure

32.—(1) Every employer who has to any extent control of any equipment or apparatus which is used in connection with a medical exposure shall, having regard to the extent of his control over the equipment, ensure that such equipment is of such design or construction and is so installed and maintained as to be capable of restricting so far as is reasonably practicable the exposure to ionising radiation of any person who is undergoing a medical exposure to the extent that this is compatible with the intended clinical purpose or research objective.

(**a**) 1993 c. 12.
(**b**) 1974 c. 37; section 6 was amended by the Consumer Protection Act 1987 (c. 43), section 36 and Schedule 3.

(2) An employer who has to any extent control of any radiation equipment which is used for the purpose of diagnosis and which is installed after the date of the coming into force of these Regulations shall, having regard to the extent of his control over the equipment, ensure that such equipment is provided, where practicable, with suitable means for informing the user of that equipment of the quantity of radiation produced by that equipment during a radiological procedure.

(3) Every employer in respect of whom a duty is imposed by paragraph (1) shall, to the extent that it is reasonable for him to do so having regard to the extent of his control over the equipment, make arrangements for a suitable quality assurance programme to be provided in respect of the equipment or apparatus for the purpose of ensuring that it remains capable of restricting so far as is reasonably practicable exposure to the extent that this is compatible with the intended clinical purpose or research objective.

(4) Without prejudice to the generality of paragraph (3), the quality assurance programme required by that paragraph shall require the carrying out of—

    (a) in respect of equipment or apparatus first used after the coming into force of this regulation, adequate testing of that equipment or apparatus before it is first used for clinical purposes;

    (b) adequate testing of the performance of the equipment or apparatus at appropriate intervals and after any major maintenance procedure to that equipment or apparatus;

    (c) where appropriate, such measurements at suitable intervals as are necessary to enable the assessment of representative doses from any radiation equipment to persons undergoing medical exposures.

(5) Every employer who has to any extent control of any radiation equipment shall take all such steps as are reasonably practicable to prevent the failure of any such equipment where such failure could result in an exposure to ionising radiation greater than that intended and to limit the consequences of any such failure.

(6) Where a radiation employer suspects or has been informed that an incident may have occurred in which a person while undergoing a medical exposure was, as the result of a malfunction of, or defect in, radiation equipment under the control of that employer, exposed to ionising radiation to an extent much greater than that intended, he shall make an immediate investigation of the suspected incident and, unless that investigation shows beyond reasonable doubt that no such incident has occurred, shall forthwith notify the Executive thereof and make or arrange for a detailed investigation of the circumstances of the exposure and an assessment of the dose received.

(7) A radiation employer who makes any investigation in accordance with paragraph (6) shall make a report of that investigation and shall—

    (a) in respect of an immediate report, keep that report or a copy thereof for a period of at least 2 years from the date on which it was made; and

    (b) in respect of a detailed report, keep that report or a copy thereof for a period of at least 50 years from the date on which it was made.

(8) In this regulation, "radiation equipment" means equipment which delivers ionising radiation to the person undergoing a medical exposure and equipment which directly controls the extent of the exposure.

Misuse of or interference with sources of ionising radiation

33. No person shall intentionally or recklessly misuse or without reasonable excuse interfere with any radioactive substance or any electrical equipment to which these Regulations apply.

PART VII

DUTIES OF EMPLOYEES AND MISCELLANEOUS

Duties of employees

34.—(1) An employee who is engaged in work with ionising radiation shall not knowingly expose himself or any other person to ionising radiation to an extent greater than is reasonably necessary for the purposes of his work, and shall exercise reasonable care while carrying out such work.

(2) Every employee who is engaged in work with ionising radiation and for whom personal protective equipment is provided pursuant to regulation 8(2)(c) shall—
- (a) make full and proper use of any such personal protective equipment;
- (b) forthwith report to his employer any defect he discovers in any such personal protective equipment; and
- (c) take all reasonable steps to ensure that any such personal protective equipment is returned after use to the accommodation provided for it.

(3) It shall be the duty of every outside worker not to misuse the radiation passbook issued to him or falsify or attempt to falsify any of the information contained in it.

(4) Any employee to whom regulation 21(1) or regulation 12(2)(b) relates shall comply with any reasonable requirement imposed on him by his employer for the purposes of making the measurements and assessments required under regulation 21(1) and regulation 23(1).

(5) An employee who is subject to medical surveillance under regulation 24 shall, when required by his employer and at the cost of the employer, present himself during his working hours for such medical examination and tests as may be required for the purposes of paragraph (2) of that regulation and shall provide the appointed doctor or employment medical adviser with such information concerning his health as the appointed doctor or employment medical adviser may reasonably require.

(6) Where an employee has reasonable cause to believe that—
- (a) he or some other person has received an overexposure;
- (b) an occurrence mentioned in paragraph (1) or (3) of regulation 30 has occurred; or
- (c) an incident mentioned in regulation 32(6) has occurred,

he shall forthwith notify his employer of that belief.

Approval of dosimetry services

35.—(1) The Executive (or such other person as may from time to time be specified in writing by the Executive) may, by a certificate in writing, approve (in accordance with such criteria as may from time to time be specified by the Executive) a suitable dosimetry service for such of the purposes of these Regulations as are specified in the certificate.

(2) A certificate made pursuant to paragraph (1) may be made subject to conditions and may be revoked in writing at any time.

(3) The Executive (or such other person as may from time to time be specified in writing by the Executive) may at such suitable periods as it considers appropriate carry out a re-assessment of any approval granted pursuant to paragraph (1).

Defence on contravention

36.—(1) In any proceedings against an employer for an offence under regulation 6(2), it shall be a defence for that employer to prove that—
- (a) he neither knew nor had reasonable cause to believe that he had carried out or might be required to carry out work subject to notification under that paragraph; and
- (b) in a case where he discovered that he had carried out or was carrying out work subject to notification under that paragraph, he had forthwith notified the Executive of the information required by that paragraph.

(2) In any proceedings against an employer for an offence under regulation 7, it shall be a defence for that employer to prove that—
- (a) he neither knew nor had reasonable cause to believe that he had commenced a new activity involving work with ionising radiation; and
- (b) in a case where he had discovered that he had commenced a new activity involving work with ionising radiation, he had as soon as practicable made an assessment as required by the said regulation 7.

(3) In any proceedings against an employer for an offence under regulation 27(2) it shall be a defence for that employer to prove that—
- (a) he had received and reasonably relied on a written undertaking from the supplier of the article concerned that it complied with the requirements of that paragraph; and
- (b) he had complied with the requirements of paragraph (3) of that regulation.

(4) In any proceedings against an employer of an outside worker for a breach of a duty under these Regulations it shall be a defence for that employer to show that—
- (a) he had entered into a contract in writing with the employer who had designated an area as a controlled area and in which the outside worker was working or was to work for that employer to perform that duty on his behalf; and
- (b) the breach of duty was a result of the failure of the employer referred to in sub-paragraph (a) above to fulfil that contract.

(5) In any proceedings against any employer who has designated a controlled area in which any outside worker is working or is to work for a breach of a duty under these Regulations it shall be a defence for that employer to show that—
- (a) he had entered into a contract in writing with the employer of an outside worker for that employer to perform that duty on his behalf; and
- (b) the breach of duty was a result of the failure of the employer referred to in sub-paragraph (a) above to fulfil that contract.

(6) The person charged shall not, without leave of the court, be entitled to rely on the defence referred to in paragraph (4) or (5) unless, within a period ending seven clear days before the hearing, he has served on the prosecutor a notice in writing that he intends to rely on the defence and this notice shall be accompanied by a copy of the contract on which he intends to rely and, if that contract is not in English, an accurate translation of that contract into English.

(7) For the purposes of enabling the other party to be charged with and convicted of an offence by virtue of section 36 of the Health and Safety at Work etc. Act 1974, a person who establishes a defence under this regulation shall nevertheless be treated for the purposes of that section as having committed the offence.

Exemption certificates

37.—(1) Subject to paragraph (2), the Executive may, by a certificate in writing, exempt—
- (a) any person or class of persons;
- (b) any premises or class of premises; or
- (c) any equipment, apparatus or substance or class of equipment, apparatus or substance,

from any requirement or prohibition imposed by these Regulations and any such exemption may be granted subject to conditions and to a limit of time and may be revoked by a certificate in writing at any time.

(2) The Executive shall not grant an exemption unless, having regard to the circumstances of the case and in particular to—
- (a) the conditions, if any, which it proposes to attach to the exemption; and
- (b) any other requirements imposed by or under any enactments which apply to the case,

it is satisfied that—
- (c) the health and safety of persons who are likely to be affected by the exemption will not be prejudiced in consequence of it; and
- (d) compliance with the fundamental radiation protection provisions underlying regulations 8(1) and (2)(a), 11, 12(1), 16(1) and 3, 19(1), 20(1), 21(1), 24(2) and 32(1) will be achieved.

Extension outside Great Britain

38.—(1) Subject to paragraph (2), these Regulations shall apply to any work outside Great Britain to which sections 1 to 59 and 80 to 82 of the Health and Safety at Work etc. Act 1974 apply by virtue of the Health and Safety at Work etc. Act 1974 (Application outside Great Britain) Order 1995(**a**) as they apply to work within Great Britain.

(2) For the purposes of paragraph (1), in any case where it is not reasonably practicable for an employer to comply with the requirements of these Regulations in so far as they relate to functions being performed by an appointed doctor or employment medical adviser or by an approved dosimetry service, it shall be sufficient compliance with any such requirements if the employer makes arrangements affording an equivalent standard of protection for his employees and those arrangements are set out in local rules.

Transitional provisions

39.—(1) Where on or before 26th February 2000 an employer commences for the first time work which is required to be notified under regulation 6(2), it shall be sufficient compliance with that regulation if the employer notifies the Executive and notifies the required particulars before 29th January 2000.

(2) A contingency plan made pursuant to the requirements of regulation 27 of the Ionising Radiations Regulations 1985(**b**) and which complied with that regulation immediately before the coming into force of these Regulations shall, for the purposes of regulation 12, be treated as if made pursuant to paragraph (1) of that regulation.

(3) A certificate of approval granted by the Executive in respect of an approved dosimetry service under regulation 15 of the Ionising Radiations Regulations 1985 and which is valid immediately before the date of the coming into force of these Regulations, shall continue in force and shall be treated as if it had been granted under regulation 35 of these Regulations.

(4) A radiation passbook approved for the purposes of the Ionising Radiations (Outside Workers) Regulations 1993(**c**) and issued prior to 30th April 2000 in respect of an outside worker employed by an employer in Great Britain and which was at that date valid shall remain valid for such time as the worker to whom the passbook relates continues to be employed by the same employer.

(5) A doctor appointed in writing by the Executive prior to the coming into force of these Regulations for the purposes of the Ionising Radiations Regulations 1985 shall, until such time as the period specified in the appointment expires or the appointment is revoked, be deemed to have been appointed for the purposes of these Regulations.

(6) Until 31st December 2004, an individual who or body which had before the coming into force of these Regulations been appointed by an employer as a radiation protection adviser for the purposes of the Ionising Radiations Regulations 1985 shall be deemed to meet the criteria of competence specified by the Executive for such advisers under these Regulations.

(7) A health record which was created prior to the coming into force of these Regulations pursuant to a requirement of the Ionising Radiations Regulations 1985 shall remain valid for a period of 12 months from the date of the last entry made in it or for such shorter period as may have been specified in that record for the validity of the last entry by an appointed doctor or employment medical adviser under those Regulations, and such record shall for that period be deemed to have been kept for the purposes of regulation 24(3).

(8) A certificate of exemption issued by the Executive pursuant to paragraph (6) of regulation 27 of the Ionising Radiations Regulations 1985 and which is valid at the coming into force of these Regulations shall continue in force until such time as it is revoked by the Executive, save that the exemption from the requirements of regulation 7 of the said 1985 Regulations shall be deemed to be an exemption from the requirements of regulation 11 of these Regulations.

(**a**) S.I. 1995/263.
(**b**) S.I. 1985/1333.
(**c**) S.I. 1993/2379.

(9) Where the Executive has reasonable cause to believe that the dose received by an employee was much greater or much less than that shown in his dose record (such record having been made and maintained in accordance with regulation 13 of the Ionising Radiations Regulations 1985) the Executive may, until 30th April 2000, approve a special entry into the dose record and in such a case the employer shall arrange for the appropriate approved dosimetry service to enter the special entry in that dose record and shall give a copy of the amended dose record to the employee to whom it relates.

Modifications relating to the Ministry of Defence etc.

40.—(1) In this regulation, any reference to—
  (a) "visiting forces" is a reference to visiting forces within the meaning of any provision of Part 1 of the Visiting Forces Act 1952(**a**); and
  (b) "headquarters or organisation" is a reference to a headquarters or organisation designated for the purposes of the International Headquarters and Defence Organisations Act 1964(**b**).

(2) The Secretary of State for Defence may, in the interests of national security, by a certificate in writing exempt—
  (a) Her Majesty's Forces;
  (b) visiting forces;
  (c) any member of a visiting force working in or attached to any headquarters or organisation; or
  (d) any person engaged in work with ionising radiation for, or on behalf of, the Secretary of State for Defence,
from all or any of the requirements or prohibitions imposed by these Regulations and any such exemption may be granted subject to conditions and to a limit of time and may be revoked at any time by a certificate in writing, except that, where any such exemption is granted, suitable arrangements shall be made for the assessment and recording of doses of ionising radiation received by persons to whom the exemption relates.

(3) Sub-paragraph (i) of regulation 21(3) shall not apply in relation to a practice carried out—
  (a) by or on behalf of the Secretary of State for Defence;
  (b) by a visiting force; or
  (c) by any member of a visiting force in or attached to any headquarters or organisation.

(4) Regulations 5 and 6 shall not apply in relation to work carried out by visiting forces or any headquarters or organisation on premises under the control of such visiting force, headquarters or organisation, as the case may be, or on premises under the control of the Secretary of State for Defence.

(5) The requirements of regulation 6 to notify the particulars specified in sub-paragraphs (d) and (e) of Schedule 2 or any of the particulars specified in Schedule 3 shall not have effect in any case where the Secretary of State for Defence decides that to do so would be against the interests of national security or where suitable alternative arrangements have been agreed with the Executive.

(6) Regulation 6(4) shall not apply to an employer in relation to work with ionising radiation undertaken for or on behalf of the Secretary of State for Defence, visiting forces or any headquarters or organisation.

(7) Regulations 22(6), (7) and (8) and regulation 24(9) shall not apply in relation to visiting forces or any member of a visiting force working in or attached to any headquarters or organisation.

(8) In regulation 25(1) the requirement to notify the Executive of a suspected overexposure and the results of the consequent investigation and assessment shall not apply in relation to the exposure of—
  (a) a member of a visiting force; or
  (b) a member of a visiting force working in or attached to a headquarters or organisation.

(**a**) 1952 c. 67.
(**b**) 1964 c. 5.

Modification, revocation and saving

41.—(1) The enactments referred to in Schedule 9 shall be modified in accordance with the provisions of that Schedule.

(2) Subject to paragraph (3), the following enactments are hereby revoked—
- (a) the Ionising Radiations Regulations 1985(**a**);
- (b) the Ionising Radiations (Outside Workers) Regulations 1993(**b**);
- (c) Part VI of Schedule 2 to the Personal Protective Equipment at Work Regulations 1992(**c**).

(3) Regulation 26 (Special hazard assessment) of the Ionising Radiations Regulations 1985 (in this paragraph referred to as "the 1985 Regulations") shall continue in force and, in respect of any employer subject to the said regulation 26, the following provisions shall also continue in force—
- (a) paragraphs (1) to (3), (4)(b) and (c) and (5) of regulation 27 (Contingency plans) with the modification that—
  - (i) in paragraph (1), the reference to regulation 25(1) of the 1985 Regulations shall be treated as a reference to regulation 7(1) or (2) of these Regulations;
  - (ii) in paragraph (1)(b), the reference to regulation 8(1) of and Schedule 6 to the 1985 Regulations shall be treated as a reference to regulation 16 of these Regulations;
  - (iii) in paragraph (4)(b), the reference to regulation 13(2) of the 1985 Regulations shall be treated as a reference to regulation 21(2) of these Regulations;
- (b) any other provisions of the 1985 Regulations in so far as is necessary to give effect to the provisions specified in this paragraph.

(4) Every register, certificate or record which was required to be kept in pursuance of any regulation revoked by paragraph (2) shall, notwithstanding that paragraph, be kept in the same manner and for the same period as if these Regulations had not been made, except that the Executive may approve the keeping of records at a place or in a form other than the place where, or the form in which, records were required to be kept under the regulation so revoked.

Signed by order of the Secretary of State.

Larry Whitty
Parliamentary Under Secretary of State,
3rd December 1999   Department of the Environment, Transport and the Regions.

(**a**) S.I. 1985/1333.
(**b**) S.I. 1993/2379.
(**c**) S.I. 1992/2966.

SCHEDULE 1

Regulations 6(1) and 13(3)

WORK NOT REQUIRED TO BE NOTIFIED UNDER REGULATION 6

1. Work with ionising radiation shall not be required to be notified in accordance with regulation 6 when the only such work being carried out is in one or more of the following categories—

  (a) where the concentration of activity per unit mass of a radioactive substance does not exceed the concentration specified in column 2 of Part I of Schedule 8;

  (b) where the quantity of radioactive substance involved does not exceed the quantity specified in column 3 of Part I of Schedule 8;

  (c) where apparatus contains radioactive substances in a quantity exceeding the values specified in sub-paragraphs (a) and (b) above provided that—

  (i) the apparatus is of a type approved by the Executive;

  (ii) the apparatus is constructed in the form of a sealed source;

  (iii) the apparatus does not under normal operating conditions cause a dose rate of more than $1\mu Svh^{-1}$ at a distance of 0.1m from any accessible surface; and

  (iv) conditions for the disposal of the apparatus have been specified by the appropriate Agency;

  (d) the operation of any electrical apparatus to which these Regulations apply other than apparatus referred to in sub-paragraph (e) below provided that—

  (i) the apparatus is of a type approved by the Executive; and

  (ii) the apparatus does not under normal operating conditions cause a dose rate of more than $1\mu Svh^{-1}$ at a distance of 0.1m from any accessible surface;

  (e) the operation of—

  (i) any cathode ray tube intended for the display of visual images; or

  (ii) any other electrical apparatus operating at a potential difference not exceeding 30kV,

  provided that the operation of the tube or apparatus does not under normal operating conditions cause a dose rate of more than $1\mu Svh^{-1}$ at a distance of 0.1m from any accessible surface;

  (f) where the work involves material contaminated with radioactive substances resulting from authorised releases which the appropriate Agency has declared not to be subject to further control.

2. In this Schedule, "the appropriate Agency" has the meaning assigned to it by section 47(1) of the Radioactive Substances Act 1993(**a**).

SCHEDULE 2

Regulation 6(2)

PARTICULARS TO BE PROVIDED IN A NOTIFICATION UNDER REGULATION 6(2)

The following particulars shall be given in a notification under regulation 6(2)—

  (a) the name and address of the employer and a contact telephone or fax number or electronic mail address;

  (b) the address of the premises where or from where the work activity is to be carried out and a telephone or fax number or electronic mail address at such premises;

  (c) the nature of the business of the employer;

  (d) into which of the following categories the source or sources of ionising radiation fall—

  (i) sealed source;

  (ii) unsealed radioactive substance;

  (iii) electrical equipment;

  (iv) an atmosphere containing the short-lived daughters of radon 222;

  (e) whether or not any source is to be used at premises other than the address given at sub-paragraph (b) above; and

  (f) dates of notification and commencement of the work activity.

(**a**) 1993 c. 12; section 47 was amended by the Environment Act 1995 (c. 25), Schedule 22, paragraph 227.

SCHEDULE 3

Regulation 6(3)

ADDITIONAL PARTICULARS THAT THE EXECUTIVE MAY REQUIRE

The following additional particulars may be required under regulation 6(3)—
- (a) a description of the work with ionising radiation;
- (b) particulars of the source or sources of ionising radiation including the type of electrical equipment used or operated and the nature of any radioactive substance;
- (c) the quantities of any radioactive substance involved in the work;
- (d) the identity of any person engaged in the work;
- (e) the date of commencement and the duration of any period over which the work is carried on;
- (f) the location and description of any premises at which the work is carried out on each occasion that it is so carried out;
- (g) the date of termination of the work;
- (h) further information on any of the particulars listed in Schedule 2.

SCHEDULE 4

Regulation 11

DOSE LIMITS

PART I

CLASSES OF PERSONS TO WHOM DOSE LIMITS APPLY

Employees of 18 years of age or above

1. For the purposes of regulation 11(1), the limit on effective dose for any employee of 18 years of age or above shall be 20 mSv in any calendar year.

2. Without prejudice to paragraph 1—
   - (a) the limit on equivalent dose for the lens of the eye shall be 150 mSv in a calendar year;
   - (b) the limit on equivalent dose for the skin shall be 500 mSv in a calendar year as applied to the dose averaged over any area of 1cm^2 regardless of the area exposed;
   - (c) the limit on equivalent dose for the hands, forearms, feet and ankles shall be 500 mSv in a calendar year.

Trainees aged under 18 years

3. For the purposes of regulation 11(1), the limit on effective dose for any trainee under 18 years of age shall be 6 mSv in any calendar year.

4. Without prejudice to paragraph 3—
   - (a) the limit on equivalent dose for the lens of the eye shall be 50 mSv in a calendar year;
   - (b) the limit on equivalent dose for the skin shall be 150 mSv in a calendar year as applied to the dose averaged over any area of 1 cm^2 regardless of the area exposed;
   - (c) the limit on equivalent dose for the hands, forearms, feet and ankles shall be 150 mSv in a calendar year.

Women of reproductive capacity

5. Without prejudice to paragraphs 1 and 3, the limit on equivalent dose for the abdomen of a women of reproductive capacity who is at work, being the equivalent dose from external radiation resulting from exposure to ionising radiation averaged throughout the abdomen, shall be 13 mSv in any consecutive period of three months.

Other persons

6. Subject to paragraph 7, for the purposes of regulation 11(1) the limit on effective dose for any person other than an employee or trainee, including any person below the age of 16, shall be 1 mSv in any calendar year.

7. Paragraph 6 shall not apply in relation to any person (not being a comforter or carer) who may be exposed to ionising radiation resulting from the medical exposure of another and in such a case the limit on effective dose for any such person shall be 5 mSv in any period of 5 consecutive calendar years.

8. Without prejudice to paragraphs 6 and 7—
   - (a) the limit on equivalent dose for the lens of the eye shall be 15 mSv in any calendar year;

(b) the limit on equivalent dose for the skin shall be 50 mSv in any calendar year averaged over any 1 cm^2 area regardless of the area exposed;

(c) the limit on equivalent dose for the hands, forearms, feet and ankles shall be 50 mSv in a calendar year.

PART II

9. For the purposes of regulation 11(2), the limit on effective dose for employees of 18 years or above shall be 100 mSv in any period of five consecutive calendar years subject to a maximum effective dose of 50 mSv in any single calendar year.

10. Without prejudice to paragraph 9—

(a) the limit on equivalent dose for the lens of the eye shall be 150 mSv in a calendar year;

(b) the limit on equivalent dose for the skin shall be 500 mSv in a calendar year as applied to the dose averaged over any area of 1cm^2 regardless of the area exposed;

(c) the limit on equivalent dose for the hands, forearms, feet and ankles shall be 500 mSv in a calendar year.

11. Without prejudice to paragraph 9, the limit on equivalent dose for the abdomen of a woman of reproductive capacity who is at work, being the equivalent dose from external radiation resulting from exposure to ionising radiation averaged throughout the abdomen, shall be 13 mSv in any consecutive period of three months.

12. The employer shall ensure that any employee in respect of whom regulation 11(2) applies is not exposed to ionising radiation to an extent that any dose limit specified in paragraphs 9 to 11 is exceeded.

13. An employer shall not put into effect a system of dose limitation in pursuance of regulation 11(2) unless—

(a) the radiation protection adviser and any employees who are affected have been consulted;

(b) any employees affected and the approved dosimetry service have been informed in writing of the decision and of the reasons for that decision; and

(c) notice has been given to the Executive at least 28 days (or such shorter period as the Executive may allow) before the decision is put into effect giving the reasons for the decision.

14. Where there is reasonable cause to believe that any employee has been exposed to an effective dose greater than 20 mSv in any calendar year, the employer shall, as soon as is practicable—

(a) undertake an investigation into the circumstances of the exposure for the purpose of determining whether the dose limit referred to in paragraph 9 is likely to be complied with; and

(b) notify the Executive of that suspected exposure.

15. An employer shall review the decision to put into effect a system of dose limitation pursuant to regulation 11(2) at appropriate intervals and in any event not less than once every five years.

16. Where as a result of a review undertaken pursuant to paragraph 15 an employer proposes to revert to a system of annual dose limitation pursuant to regulation 11(1), the provisions of paragraph 13 shall apply as if the reference in that paragraph to regulation 11(2) was a reference to regulation 11(1).

17. Where an employer puts into effect a system of dose limitation in pursuance of regulation 11(2), he shall record the reasons for that decision and shall ensure that the record is preserved for a period of 50 years from the date of its making.

18. In any case where—

(a) the dose limits specified in paragraph 9 are being applied by a radiation employer in respect of an employee; and

(b) the Executive is not satisfied that it is impracticable for that employee to be subject to the dose limit specified in paragraph 1 of Part I of this Schedule,

the Executive may require the employer to apply the dose limit specified in paragraph 1 of Part I with effect from such time as the Executive may consider appropriate having regard to the interests of the employee concerned.

19. In any case where, as a result of a review undertaken pursuant to paragraph 15, an employer proposes to revert to an annual dose limitation pursuant to regulation 11(2), the Executive may require the employer to defer the implementation of that decision to such time as the Executive may consider appropriate having regard to the interests of the employee concerned.

20. Any person who is aggrieved by the decision of the Executive taken pursuant to paragraphs 18 or 19 may appeal to the Secretary of State.

21. Sub-sections (2) to (6) of section 44 of the 1974 Act shall apply for the purposes of paragraph 20 as they apply to an appeal under section 44(1) of that Act.

22. The Health and Safety Licensing Appeals (Hearings Procedure) Rules 1974(**a**), as respects England and Wales, and the Health and Safety Licensing Appeals (Hearing Procedure) (Scotland) Rules 1974(**b**), as respects Scotland, shall apply to an appeal under paragraph 20 as they apply to an appeal under sub-section (1) of the said section 44, but with the modification that references to a licensing authority are to be read as references to the Executive.

SCHEDULE 5

Regulation 13(1)

MATTERS IN RESPECT OF WHICH A RADIATION PROTECTION ADVISER MUST BE CONSULTED BY A RADIATION EMPLOYER

1. The implementation of requirements as to controlled and supervised areas.

2. The prior examination of plans for installations and the acceptance into service of new or modified sources of ionising radiation in relation to any engineering controls, design features, safety features and warning devices provided to restrict exposure to ionising radiation.

3. The regular calibration of equipment provided for monitoring levels of ionising radiation and the regular checking that such equipment is serviceable and correctly used.

4. The periodic examination and testing of engineering controls, design features, safety features and warning devices and regular checking of systems of work provided to restrict exposure to ionising radiation.

SCHEDULE 6

Regulation 21(5)

PARTICULARS TO BE ENTERED IN THE RADIATION PASSBOOK

1. Individual serial number of the passbook.

2. A statement that the passbook has been approved by the Executive for the purpose of these Regulations.

3. Date of issue of the passbook by the approved dosimetry service.

4. The name, telephone number and mark of endorsement of the issuing approved dosimetry service.

5. The name, address, telephone and telex/fax number of the employer.

6. Full name (surname, forenames), date of birth, gender and national insurance number of the outside worker to whom the passbook has been issued.

7. Date of the last medical review of the outside worker and the relevant classification in the health record maintained under regulation 24 as fit, fit subject to conditions (which shall be specified) or unfit.

8. The relevant dose limits applicable to the outside worker to whom the passbook has been issued.

9. The cumulative dose assessment in mSv for the year to date for the outside worker, external (whole body, organ or tissue) and/or internal as appropriate and the date of the end of the last assessment period.

10. In respect of services performed by the outside worker—
   (a) the name and address of the employer responsible for the controlled area;
   (b) the period covered by the performance of the services;
   (c) estimated dose information, which shall be, as appropriate—
      (i) an estimate of any whole body effective dose in mSv received by the outside worker;
      (ii) in the event of non-uniform exposure, an estimate of the equivalent dose in mSv to organs and tissues as appropriate; and
      (iii) in the event of internal contamination, an estimate of the activity taken in or the committed dose.

(**a**) S.I. 1974/2040.
(**b**) S.I. 1974/2068.

SCHEDULE 7

Regulation 24(3)

PARTICULARS TO BE CONTAINED IN A HEALTH RECORD

The following particulars shall be contained in a health record made for the purposes of regulation 24(3)—

(a) the employee's—
   (i) full name;
   (ii) sex;
   (iii) date of birth;
   (iv) permanent address; and
   (v) National Insurance number;

(b) the date of the employee's commencement as a classified person in present employment;

(c) the nature of the employee's employment;

(d) in the case of a female employee, a statement as to whether she is likely to receive in any consecutive period of three months an equivalent dose of ionising radiation for the abdomen exceeding 13 mSv;

(e) the date of last medical examination or health review carried out in respect of the employee;

(f) the type of the last medical examination or health review carried out in respect of the employee;

(g) a statement by the appointed doctor or employment medical adviser made as a result of the last medical examination or health review carried out in respect of the employee classifying the employee as fit, fit subject to conditions (which should be specified) or unfit;

(h) in the case of a female employee in respect of whom a statement has been made under paragraph (d) to the effect that she is likely to receive in any consecutive period of three months an equivalent dose of ionising radiation for the abdomen exceeding 13 mSv, a statement by the appointed doctor or employment medical adviser certifying whether in his professional opinion the employee should be subject to the additional dose limit specified in paragraphs 5 and 11 of Schedule 4;

(i) in relation to each medical examination and health review, the name and signature of the appointed doctor or employment medical adviser;

(j) the name and address of the approved dosimetry service with whom arrangements have been made for maintaining the dose record in accordance with regulation 21.

SCHEDULE 8

Regulation 2(4) and 30(1) and (2) and Schedule 1

QUANTITIES AND CONCENTRATIONS OF RADIONUCLIDES

PART I

TABLE OF RADIONUCLIDES

1	2	3	4	5
Radionuclide name, symbol, isotope	**Concentration for notification. Regulation 6 and Schedule 1 (Bq/g)**	**Quantity for notification. Regulation 6 and Schedule 1 (Bq)**	**Quantity for notification of occurrences. Regulation 30(1) (Bq)**	**Quantity for notification of occurrences. Regulation 30(3) (Bq)**
Hydrogen				
Tritiated Compounds	$1\ 10^6$	$1\ 10^9$	$1\ 10^{12}$	$1\ 10^{10}$
Elemental	$1\ 10^6$	$1\ 10^9$	$1\ 10^{13}$	$1\ 10^{10}$
Beryllium				
Be-7	$1\ 10^3$	$1\ 10^7$	$1\ 10^{12}$	$1\ 10^8$
Be-10	$1\ 10^4$	$1\ 10^6$	$1\ 10^{10}$	$1\ 10^7$
Carbon				
C-11	$1\ 10^1$	$1\ 10^6$	$1\ 10^{13}$	$1\ 10^7$
C-11 monoxide	$1\ 10^1$	$1\ 10^9$	$1\ 10^{12}$	$1\ 10^{10}$
C-11 dioxide	$1\ 10^1$	$1\ 10^9$	$1\ 10^{12}$	$1\ 10^{10}$
C-14	$1\ 10^4$	$1\ 10^7$	$1\ 10^{11}$	$1\ 10^8$
C-14 monoxide	$1\ 10^8$	$1\ 10^{11}$	$1\ 10^{14}$	$1\ 10^{12}$
C-14 dioxide	$1\ 10^7$	$1\ 10^{11}$	$1\ 10^{13}$	$1\ 10^{12}$
Nitrogen				
N-13	$1\ 10^2$	$1\ 10^9$	$1\ 10^9$	
Oxygen				
O-15	$1\ 10^2$	$1\ 10^9$	$1\ 10^{10}$	
Fluorine				
F-18	$1\ 10^1$	$1\ 10^6$	$1\ 10^{13}$	$1\ 10^7$
Neon				
Ne-19	$1\ 10^2$	$1\ 10^9$	$1\ 10^9$	

1	2	3	4	5
Radionuclide name, symbol, isotope	Concentration for notification. Regulation 6 and Schedule 1 (Bq/g)	Quantity for notification. Regulation 6 and Schedule 1 (Bq)	Quantity for notification of occurrences. Regulation 30(1) (Bq)	Quantity for notification of occurrences. Regulation 30(3) (Bq)
Sodium				
Na-22	$1\ 10^1$	$1\ 10^6$	$1\ 10^{10}$	$1\ 10^7$
Na-24	$1\ 10^1$	$1\ 10^5$	$1\ 10^{11}$	$1\ 10^6$
Magnesium				
Mg-28+	$1\ 10^1$	$1\ 10^5$	$1\ 10^{11}$	$1\ 10^6$
Aluminium				
Al-26	$1\ 10^1$	$1\ 10^5$	$1\ 10^{10}$	$1\ 10^6$
Silicon				
Si-31	$1\ 10^3$	$1\ 10^6$	$1\ 10^{13}$	$1\ 10^7$
Si-32	$1\ 10^3$	$1\ 10^6$	$1\ 10^9$	$1\ 10^7$
Phosphorus				
P-32	$1\ 10^3$	$1\ 10^5$	$1\ 10^{10}$	$1\ 10^6$
P-33	$1\ 10^5$	$1\ 10^8$	$1\ 10^{11}$	$1\ 10^9$
Sulphur				
S-35	$1\ 10^5$	$1\ 10^8$	$1\ 10^{11}$	$1\ 10^9$
S-35 (organic)	$1\ 10^5$	$1\ 10^8$	$1\ 10^{12}$	$1\ 10^9$
S-35 Vapour	$1\ 10^6$	$1\ 10^9$	$1\ 10^{12}$	
Chlorine				
Cl-36	$1\ 10^4$	$1\ 10^6$	$1\ 10^{10}$	$1\ 10^7$
Cl-38	$1\ 10^1$	$1\ 10^5$	$1\ 10^{13}$	$1\ 10^6$
Cl-39	$1\ 10^1$	$1\ 10^5$	$1\ 10^{13}$	$1\ 10^6$
Argon				
Ar-37	$1\ 10^6$	$1\ 10^8$	$1\ 10^{13}$	
Ar-39	$1\ 10^7$	$1\ 10^4$	$1\ 10^{12}$	
Ar-41	$1\ 10^2$	$1\ 10^9$	$1\ 10^9$	
Potassium				
K-40	$1\ 10^2$	$1\ 10^6$	$1\ 10^{10}$	$1\ 10^7$
K-42	$1\ 10^2$	$1\ 10^6$	$1\ 10^{12}$	$1\ 10^7$

1	2	3	4	5
Radionuclide name, symbol, isotope	Concentration for notification. Regulation 6 and Schedule 1 (Bq/g)	Quantity for notification. Regulation 6 and Schedule 1 (Bq)	Quantity for notification of occurrences. Regulation 30(1) (Bq)	Quantity for notification of occurrences. Regulation 30(3) (Bq)
K-43	$1\ 10^1$	$1\ 10^6$	$1\ 10^{11}$	$1\ 10^7$
K-44	$1\ 10^1$	$1\ 10^5$	$1\ 10^{13}$	$1\ 10^6$
K-45	$1\ 10^1$	$1\ 10^5$	$1\ 10^{13}$	$1\ 10^6$
Calcium				
Ca-41	$1\ 10^5$	$1\ 10^7$	$1\ 10^{12}$	$1\ 10^8$
Ca-45	$1\ 10^4$	$1\ 10^7$	$1\ 10^{10}$	$1\ 10^8$
Ca-47	$1\ 10^1$	$1\ 10^6$	$1\ 10^{11}$	$1\ 10^7$
Scandium				
Sc-43	$1\ 10^1$	$1\ 10^6$	$1\ 10^{12}$	$1\ 10^7$
Sc-44	$1\ 10^1$	$1\ 10^5$	$1\ 10^{12}$	$1\ 10^6$
Sc-44m	$1\ 10^2$	$1\ 10^7$	$1\ 10^{11}$	$1\ 10^8$
Sc-46	$1\ 10^1$	$1\ 10^6$	$1\ 10^{10}$	$1\ 10^7$
Sc-47	$1\ 10^2$	$1\ 10^6$	$1\ 10^{11}$	$1\ 10^7$
Sc-48	$1\ 10^1$	$1\ 10^5$	$1\ 10^{11}$	$1\ 10^6$
Sc-49	$1\ 10^3$	$1\ 10^5$	$1\ 10^{14}$	$1\ 10^6$
Titanium				
Ti-44+	$1\ 10^1$	$1\ 10^5$	$1\ 10^9$	$1\ 10^6$
Ti-45	$1\ 10^1$	$1\ 10^6$	$1\ 10^{12}$	$1\ 10^7$
Vanadium				
V-47	$1\ 10^1$	$1\ 10^5$	$1\ 10^{13}$	$1\ 10^6$
V-48	$1\ 10^1$	$1\ 10^5$	$1\ 10^{10}$	$1\ 10^6$
V-49	$1\ 10^4$	$1\ 10^7$	$1\ 10^{12}$	$1\ 10^8$
Chromium				
Cr-48	$1\ 10^2$	$1\ 10^6$	$1\ 10^{12}$	$1\ 10^7$
Cr-49	$1\ 10^1$	$1\ 10^6$	$1\ 10^{13}$	$1\ 10^7$
Cr-51	$1\ 10^3$	$1\ 10^7$	$1\ 10^{12}$	$1\ 10^8$
Manganese				
Mn-51	$1\ 10^1$	$1\ 10^5$	$1\ 10^{13}$	$1\ 10^6$
Mn-52	$1\ 10^1$	$1\ 10^5$	$1\ 10^{10}$	$1\ 10^6$
Mn-52m	$1\ 10^1$	$1\ 10^5$	$1\ 10^{13}$	$1\ 10^6$

1	2	3	4	5
Radionuclide name, symbol, isotope	Concentration for notification. Regulation 6 and Schedule 1 (Bq/g)	Quantity for notification. Regulation 6 and Schedule 1 (Bq)	Quantity for notification of occurrences. Regulation 30(1) (Bq)	Quantity for notification of occurrences. Regulation 30(3) (Bq)
Mn-53	$1\ 10^4$	$1\ 10^9$	$1\ 10^{12}$	$1\ 10^{10}$
Mn-54	$1\ 10^1$	$1\ 10^6$	$1\ 10^{11}$	$1\ 10^7$
Mn-56	$1\ 10^1$	$1\ 10^5$	$1\ 10^{12}$	$1\ 10^6$
Iron				
Fe-52	$1\ 10^1$	$1\ 10^6$	$1\ 10^{12}$	$1\ 10^7$
Fe-55	$1\ 10^4$	$1\ 10^6$	$1\ 10^{11}$	$1\ 10^7$
Fe-59	$1\ 10^1$	$1\ 10^6$	$1\ 10^{10}$	$1\ 10^7$
Fe-60+	$1\ 10^2$	$1\ 10^5$	$1\ 10^8$	$1\ 10^6$
Cobalt				
Co-55	$1\ 10^1$	$1\ 10^6$	$1\ 10^{11}$	$1\ 10^7$
Co-56	$1\ 10^1$	$1\ 10^5$	$1\ 10^{10}$	$1\ 10^6$
Co-57	$1\ 10^2$	$1\ 10^6$	$1\ 10^{11}$	$1\ 10^7$
Co-58	$1\ 10^1$	$1\ 10^6$	$1\ 10^{10}$	$1\ 10^7$
Co-58m	$1\ 10^4$	$1\ 10^7$	$1\ 10^{13}$	$1\ 10^8$
Co-60	$1\ 10^1$	$1\ 10^5$	$1\ 10^{10}$	$1\ 10^6$
Co-60m	$1\ 10^3$	$1\ 10^6$	$1\ 10^{16}$	$1\ 10^7$
Co-61	$1\ 10^2$	$1\ 10^6$	$1\ 10^{13}$	$1\ 10^7$
Co-62m	$1\ 10^1$	$1\ 10^5$	$1\ 10^{13}$	$1\ 10^6$
Nickel				
Ni-56	$1\ 10^1$	$1\ 10^6$	$1\ 10^{11}$	$1\ 10^7$
Ni-57	$1\ 10^1$	$1\ 10^6$	$1\ 10^{11}$	$1\ 10^7$
Ni-59	$1\ 10^4$	$1\ 10^8$	$1\ 10^{11}$	$1\ 10^9$
Ni-63	$1\ 10^5$	$1\ 10^8$	$1\ 10^{11}$	$1\ 10^9$
Ni-65	$1\ 10^1$	$1\ 10^6$	$1\ 10^{13}$	$1\ 10^7$
Ni-66	$1\ 10^4$	$1\ 10^7$	$1\ 10^{11}$	$1\ 10^8$
Copper				
Cu-60	$1\ 10^1$	$1\ 10^5$	$1\ 10^{13}$	$1\ 10^6$
Cu-61	$1\ 10^1$	$1\ 10^6$	$1\ 10^{12}$	$1\ 10^7$
Cu-64	$1\ 10^2$	$1\ 10^6$	$1\ 10^{12}$	$1\ 10^7$
Cu-67	$1\ 10^2$	$1\ 10^6$	$1\ 10^{11}$	$1\ 10^7$

1	2	3	4	5
Radionuclide name, symbol, isotope	Concentration for notification. Regulation 6 and Schedule 1 (Bq/g)	Quantity for notification. Regulation 6 and Schedule 1 (Bq)	Quantity for notification of occurrences. Regulation 30(1) (Bq)	Quantity for notification of occurrences. Regulation 30(3) (Bq)
Zinc				
Zn-62	$1\ 10^2$	$1\ 10^6$	$1\ 10^{12}$	$1\ 10^7$
Zn-63	$1\ 10^1$	$1\ 10^5$	$1\ 10^{13}$	$1\ 10^6$
Zn-65	$1\ 10^1$	$1\ 10^6$	$1\ 10^{10}$	$1\ 10^7$
Zn-69	$1\ 10^4$	$1\ 10^6$	$1\ 10^{14}$	$1\ 10^7$
Zn-69m	$1\ 10^2$	$1\ 10^6$	$1\ 10^{12}$	$1\ 10^7$
Zn-71m	$1\ 10^1$	$1\ 10^6$	$1\ 10^{12}$	$1\ 10^7$
Zn-72	$1\ 10^2$	$1\ 10^6$	$1\ 10^{11}$	$1\ 10^7$
Gallium				
Ga-65	$1\ 10^1$	$1\ 10^5$	$1\ 10^{13}$	$1\ 10^6$
Ga-66	$1\ 10^1$	$1\ 10^5$	$1\ 10^{11}$	$1\ 10^6$
Ga-67	$1\ 10^2$	$1\ 10^6$	$1\ 10^{11}$	$1\ 10^7$
Ga-68	$1\ 10^1$	$1\ 10^5$	$1\ 10^{13}$	$1\ 10^6$
Ga-70	$1\ 10^3$	$1\ 10^6$	$1\ 10^{14}$	$1\ 10^7$
Ga-72	$1\ 10^1$	$1\ 10^5$	$1\ 10^{11}$	$1\ 10^6$
Ga-73	$1\ 10^2$	$1\ 10^6$	$1\ 10^{12}$	$1\ 10^7$
Germanium				
Ge-66	$1\ 10^1$	$1\ 10^6$	$1\ 10^{13}$	$1\ 10^7$
Ge-67	$1\ 10^1$	$1\ 10^5$	$1\ 10^{13}$	$1\ 10^6$
Ge-68 +	$1\ 10^1$	$1\ 10^5$	$1\ 10^{10}$	$1\ 10^6$
Ge-69	$1\ 10^1$	$1\ 10^6$	$1\ 10^{11}$	$1\ 10^7$
Ge-71	$1\ 10^4$	$1\ 10^8$	$1\ 10^{13}$	$1\ 10^9$
Ge-75	$1\ 10^3$	$1\ 10^6$	$1\ 10^{14}$	$1\ 10^7$
Ge-77	$1\ 10^1$	$1\ 10^5$	$1\ 10^{12}$	$1\ 10^6$
Ge-78	$1\ 10^2$	$1\ 10^6$	$1\ 10^{13}$	$1\ 10^7$
Arsenic				
As-69	$1\ 10^1$	$1\ 10^5$	$1\ 10^{13}$	$1\ 10^6$
As-70	$1\ 10^1$	$1\ 10^5$	$1\ 10^{12}$	$1\ 10^6$
As-71	$1\ 10^1$	$1\ 10^6$	$1\ 10^{11}$	$1\ 10^7$
As-72	$1\ 10^1$	$1\ 10^5$	$1\ 10^{11}$	$1\ 10^6$
As-73	$1\ 10^3$	$1\ 10^7$	$1\ 10^{11}$	$1\ 10^8$
As-74	$1\ 10^1$	$1\ 10^6$	$1\ 10^{11}$	$1\ 10^7$

1	2	3	4	5
Radionuclide name, symbol, isotope	Concentration for notification. Regulation 6 and Schedule 1 (Bq/g)	Quantity for notification. Regulation 6 and Schedule 1 (Bq)	Quantity for notification of occurrences. Regulation 30(1) (Bq)	Quantity for notification of occurrences. Regulation 30(3) (Bq)
As-76	$1\ 10^2$	$1\ 10^5$	$1\ 10^{11}$	$1\ 10^6$
As-77	$1\ 10^3$	$1\ 10^6$	$1\ 10^{12}$	$1\ 10^7$
As-78	$1\ 10^1$	$1\ 10^5$	$1\ 10^{13}$	$1\ 10^6$
Selenium				
Se-70	$1\ 10^1$	$1\ 10^6$	$1\ 10^{13}$	$1\ 10^7$
Se-73	$1\ 10^1$	$1\ 10^6$	$1\ 10^{12}$	$1\ 10^7$
Se-73m	$1\ 10^2$	$1\ 10^6$	$1\ 10^{14}$	$1\ 10^7$
Se-75	$1\ 10^2$	$1\ 10^6$	$1\ 10^{11}$	$1\ 10^7$
Se-79	$1\ 10^4$	$1\ 10^7$	$1\ 10^{10}$	$1\ 10^8$
Se-81	$1\ 10^3$	$1\ 10^6$	$1\ 10^{14}$	$1\ 10^7$
Se-81m	$1\ 10^3$	$1\ 10^7$	$1\ 10^{14}$	$1\ 10^8$
Se-83	$1\ 10^1$	$1\ 10^5$	$1\ 10^{13}$	$1\ 10^6$
Bromine				
Br-74	$1\ 10^1$	$1\ 10^5$	$1\ 10^{13}$	$1\ 10^6$
Br-74m	$1\ 10^1$	$1\ 10^5$	$1\ 10^{12}$	$1\ 10^6$
Br-75	$1\ 10^1$	$1\ 10^6$	$1\ 10^{13}$	$1\ 10^7$
Br-76	$1\ 10^1$	$1\ 10^5$	$1\ 10^{11}$	$1\ 10^6$
Br-77	$1\ 10^2$	$1\ 10^6$	$1\ 10^{12}$	$1\ 10^7$
Br-80	$1\ 10^2$	$1\ 10^5$	$1\ 10^{14}$	$1\ 10^6$
Br-80m	$1\ 10^3$	$1\ 10^7$	$1\ 10^{13}$	$1\ 10^8$
Br-82	$1\ 10^1$	$1\ 10^6$	$1\ 10^{11}$	$1\ 10^7$
Br-83	$1\ 10^3$	$1\ 10^6$	$1\ 10^{13}$	$1\ 10^7$
Br-84	$1\ 10^1$	$1\ 10^5$	$1\ 10^{13}$	$1\ 10^6$
Krypton				
Kr-74	$1\ 10^2$	$1\ 10^9$	$1\ 10^9$	
Kr-76	$1\ 10^2$	$1\ 10^9$	$1\ 10^{10}$	
Kr-77	$1\ 10^2$	$1\ 10^9$	$1\ 10^9$	
Kr-79	$1\ 10^3$	$1\ 10^5$	$1\ 10^{10}$	
Kr-81	$1\ 10^4$	$1\ 10^7$	$1\ 10^{11}$	
Kr-81m	$1\ 10^3$	$1\ 10^{10}$	$1\ 10^{10}$	
Kr-83m	$1\ 10^5$	$1\ 10^{12}$	$1\ 10^{12}$	
Kr-85	$1\ 10^5$	$1\ 10^4$	$1\ 10^{12}$	
Kr-85m	$1\ 10^3$	$1\ 10^{10}$	$1\ 10^{10}$	

1	2	3	4	5
Radionuclide name, symbol, isotope	Concentration for notification. Regulation 6 and Schedule 1 (Bq/g)	Quantity for notification. Regulation 6 and Schedule 1 (Bq)	Quantity for notification of occurrences. Regulation 30(1) (Bq)	Quantity for notification of occurrences. Regulation 30(3) (Bq)
Kr-87	$1\ 10^2$	$1\ 10^9$	$1\ 10^9$	
Kr-88	$1\ 10^2$	$1\ 10^9$	$1\ 10^9$	
Rubidium				
Rb-79	$1\ 10^1$	$1\ 10^5$	$1\ 10^{13}$	$1\ 10^6$
Rb-81	$1\ 10^1$	$1\ 10^6$	$1\ 10^{12}$	$1\ 10^7$
Rb-81m	$1\ 10^3$	$1\ 10^7$	$1\ 10^{15}$	$1\ 10^8$
Rb-82m	$1\ 10^1$	$1\ 10^6$	$1\ 10^{12}$	$1\ 10^7$
Rb-83+	$1\ 10^2$	$1\ 10^6$	$1\ 10^{11}$	$1\ 10^7$
Rb-84	$1\ 10^1$	$1\ 10^6$	$1\ 10^{11}$	$1\ 10^7$
Rb-86	$1\ 10^2$	$1\ 10^5$	$1\ 10^{11}$	$1\ 10^6$
Rb-87	$1\ 10^4$	$1\ 10^7$	$1\ 10^{11}$	$1\ 10^8$
Rb-88	$1\ 10^1$	$1\ 10^5$	$1\ 10^{14}$	$1\ 10^6$
Rb-89	$1\ 10^1$	$1\ 10^5$	$1\ 10^{13}$	$1\ 10^6$
Strontium				
Sr-80	$1\ 10^3$	$1\ 10^7$	$1\ 10^{13}$	$1\ 10^8$
Sr-81	$1\ 10^1$	$1\ 10^5$	$1\ 10^{13}$	$1\ 10^6$
Sr-82+	$1\ 10^1$	$1\ 10^5$	$1\ 10^{10}$	$1\ 10^6$
Sr-83	$1\ 10^1$	$1\ 10^6$	$1\ 10^{11}$	$1\ 10^7$
Sr-85	$1\ 10^2$	$1\ 10^6$	$1\ 10^{11}$	$1\ 10^7$
Sr-85m	$1\ 10^2$	$1\ 10^7$	$1\ 10^{13}$	$1\ 10^8$
Sr-87m	$1\ 10^2$	$1\ 10^6$	$1\ 10^{13}$	$1\ 10^7$
Sr-89	$1\ 10^3$	$1\ 10^6$	$1\ 10^{10}$	$1\ 10^7$
Sr-90+	$1\ 10^2$	$1\ 10^4$	$1\ 10^9$	$1\ 10^5$
Sr-91	$1\ 10^1$	$1\ 10^5$	$1\ 10^{12}$	$1\ 10^6$
Sr-92	$1\ 10^1$	$1\ 10^6$	$1\ 10^{12}$	$1\ 10^7$
Yttrium				
Y-86	$1\ 10^1$	$1\ 10^5$	$1\ 10^{11}$	$1\ 10^6$
Y-86m	$1\ 10^2$	$1\ 10^7$	$1\ 10^{14}$	$1\ 10^8$
Y-87+	$1\ 10^1$	$1\ 10^6$	$1\ 10^{11}$	$1\ 10^7$
Y-88	$1\ 10^1$	$1\ 10^6$	$1\ 10^{10}$	$1\ 10^7$
Y-90	$1\ 10^3$	$1\ 10^5$	$1\ 10^{11}$	$1\ 10^6$
Y-90m	$1\ 10^1$	$1\ 10^6$	$1\ 10^{12}$	$1\ 10^7$
Y-91	$1\ 10^3$	$1\ 10^6$	$1\ 10^{10}$	$1\ 10^7$

1	2	3	4	5
Radionuclide name, symbol, isotope	Concentration for notification. Regulation 6 and Schedule 1 (Bq/g)	Quantity for notification. Regulation 6 and Schedule 1 (Bq)	Quantity for notification of occurrences. Regulation 30(1) (Bq)	Quantity for notification of occurrences. Regulation 30(3) (Bq)
Y-91m	$1\ 10^2$	$1\ 10^6$	$1\ 10^{13}$	$1\ 10^7$
Y-92	$1\ 10^2$	$1\ 10^5$	$1\ 10^{12}$	$1\ 10^6$
Y-93	$1\ 10^2$	$1\ 10^5$	$1\ 10^{12}$	$1\ 10^6$
Y-94	$1\ 10^1$	$1\ 10^5$	$1\ 10^{13}$	$1\ 10^6$
Y-95	$1\ 10^1$	$1\ 10^5$	$1\ 10^{14}$	$1\ 10^6$
Zirconium				
Zr-86	$1\ 10^2$	$1\ 10^7$	$1\ 10^{12}$	$1\ 10^8$
Zr-88	$1\ 10^2$	$1\ 10^6$	$1\ 10^{10}$	$1\ 10^7$
Zr-89	$1\ 10^1$	$1\ 10^6$	$1\ 10^{11}$	$1\ 10^7$
Zr-93+	$1\ 10^3$	$1\ 10^7$	$1\ 10^9$	$1\ 10^8$
Zr-95	$1\ 10^1$	$1\ 10^6$	$1\ 10^{10}$	$1\ 10^7$
Zr-97+	$1\ 10^1$	$1\ 10^5$	$1\ 10^{11}$	$1\ 10^6$
Niobium				
Nb-88	$1\ 10^1$	$1\ 10^5$	$1\ 10^{13}$	$1\ 10^6$
Nb-89 (2.03 hours)	$1\ 10^1$	$1\ 10^5$	$1\ 10^{12}$	$1\ 10^6$
Nb-89 (1.01 hour)	$1\ 10^1$	$1\ 10^5$	$1\ 10^{13}$	$1\ 10^6$
Nb-90	$1\ 10^1$	$1\ 10^5$	$1\ 10^{11}$	$1\ 10^6$
Nb-93m	$1\ 10^4$	$1\ 10^7$	$1\ 10^{11}$	$1\ 10^8$
Nb-94	$1\ 10^1$	$1\ 10^6$	$1\ 10^9$	$1\ 10^7$
Nb-95	$1\ 10^1$	$1\ 10^6$	$1\ 10^{11}$	$1\ 10^7$
Nb-95m	$1\ 10^2$	$1\ 10^7$	$1\ 10^{11}$	$1\ 10^8$
Nb-96	$1\ 10^1$	$1\ 10^5$	$1\ 10^{11}$	$1\ 10^6$
Nb-97	$1\ 10^1$	$1\ 10^6$	$1\ 10^{13}$	$1\ 10^7$
Nb-98	$1\ 10^1$	$1\ 10^5$	$1\ 10^{13}$	$1\ 10^6$
Molybdenum				
Mo-90	$1\ 10^1$	$1\ 10^6$	$1\ 10^{12}$	$1\ 10^7$
Mo-93	$1\ 10^3$	$1\ 10^8$	$1\ 10^{11}$	$1\ 10^9$
Mo-93m	$1\ 10^1$	$1\ 10^6$	$1\ 10^{12}$	$1\ 10^7$
Mo-99	$1\ 10^2$	$1\ 10^6$	$1\ 10^{11}$	$1\ 10^7$
Mo-101	$1\ 10^1$	$1\ 10^6$	$1\ 10^{13}$	$1\ 10^7$
Technetium				
Tc-93	$1\ 10^1$	$1\ 10^6$	$1\ 10^{12}$	$1\ 10^7$

1	2	3	4	5
Radionuclide name, symbol, isotope	Concentration for notification. Regulation 6 and Schedule 1 (Bq/g)	Quantity for notification. Regulation 6 and Schedule 1 (Bq)	Quantity for notification of occurrences. Regulation 30(1) (Bq)	Quantity for notification of occurrences. Regulation 30(3) (Bq)
Tc-93m	$1\ 10^1$	$1\ 10^6$	$1\ 10^{13}$	$1\ 10^7$
Tc-94	$1\ 10^1$	$1\ 10^6$	$1\ 10^{12}$	$1\ 10^7$
Tc-94m	$1\ 10^1$	$1\ 10^5$	$1\ 10^{13}$	$1\ 10^6$
Tc-95	$1\ 10^1$	$1\ 10^6$	$1\ 10^{12}$	$1\ 10^7$
Tc-95m+	$1\ 10^1$	$1\ 10^6$	$1\ 10^{11}$	$1\ 10^7$
Tc-96	$1\ 10^1$	$1\ 10^6$	$1\ 10^{11}$	$1\ 10^7$
Tc-96m	$1\ 10^3$	$1\ 10^7$	$1\ 10^{14}$	$1\ 10^8$
Tc-97	$1\ 10^3$	$1\ 10^8$	$1\ 10^{12}$	$1\ 10^9$
Tc-97m	$1\ 10^3$	$1\ 10^7$	$1\ 10^{10}$	$1\ 10^8$
Tc-98	$1\ 10^1$	$1\ 10^6$	$1\ 10^{10}$	$1\ 10^7$
Tc-99	$1\ 10^4$	$1\ 10^7$	$1\ 10^{10}$	$1\ 10^8$
Tc-99m	$1\ 10^2$	$1\ 10^7$	$1\ 10^{13}$	$1\ 10^8$
Tc-101	$1\ 10^2$	$1\ 10^6$	$1\ 10^{14}$	$1\ 10^7$
Tc-104	$1\ 10^1$	$1\ 10^5$	$1\ 10^{13}$	$1\ 10^6$
Ruthenium				
Ru-94	$1\ 10^2$	$1\ 10^6$	$1\ 10^{13}$	$1\ 10^7$
Ru-97	$1\ 10^2$	$1\ 10^7$	$1\ 10^{12}$	$1\ 10^8$
Ru-103	$1\ 10^2$	$1\ 10^6$	$1\ 10^{10}$	$1\ 10^7$
Ru-105	$1\ 10^1$	$1\ 10^6$	$1\ 10^{12}$	$1\ 10^7$
Ru-160+	$1\ 10^2$	$1\ 10^5$	$1\ 10^9$	$1\ 10^6$
Rhodium				
Rh-99	$1\ 10^1$	$1\ 10^6$	$1\ 10^{11}$	$1\ 10^7$
Rh-99m	$1\ 10^1$	$1\ 10^6$	$1\ 10^{12}$	$1\ 10^7$
Rh-100	$1\ 10^1$	$1\ 10^6$	$1\ 10^{11}$	$1\ 10^7$
Rh-101	$1\ 10^2$	$1\ 10^7$	$1\ 10^{10}$	$1\ 10^8$
Rh-101m	$1\ 10^2$	$1\ 10^7$	$1\ 10^{11}$	$1\ 10^8$
Rh-102	$1\ 10^1$	$1\ 10^6$	$1\ 10^{10}$	$1\ 10^7$
Rh-102m	$1\ 10^2$	$1\ 10^6$	$1\ 10^{10}$	$1\ 10^7$
Rh-103m	$1\ 10^4$	$1\ 10^8$	$1\ 10^{15}$	$1\ 10^9$
Rh-105	$1\ 10^2$	$1\ 10^7$	$1\ 10^{12}$	$1\ 10^8$
Rh-106m	$1\ 10^1$	$1\ 10^5$	$1\ 10^{12}$	$1\ 10^6$
Rh-107	$1\ 10^2$	$1\ 10^6$	$1\ 10^{14}$	$1\ 10^7$

1	2	3	4	5
Radionuclide name, symbol, isotope	Concentration for notification. Regulation 6 and Schedule 1 (Bq/g)	Quantity for notification. Regulation 6 and Schedule 1 (Bq)	Quantity for notification of occurrences. Regulation 30(1) (Bq)	Quantity for notification of occurrences. Regulation 30(3) (Bq)
Palladium				
Pd-100	$1\ 10^2$	$1\ 10^7$	$1\ 10^{11}$	$1\ 10^8$
Pd-101	$1\ 10^2$	$1\ 10^6$	$1\ 10^{12}$	$1\ 10^7$
Pd-103	$1\ 10^3$	$1\ 10^8$	$1\ 10^{11}$	$1\ 10^9$
Pd-107	$1\ 10^5$	$1\ 10^8$	$1\ 10^{11}$	$1\ 10^9$
Pd-109	$1\ 10^3$	$1\ 10^6$	$1\ 10^{12}$	$1\ 10^7$
Silver				
Ag-102	$1\ 10^1$	$1\ 10^5$	$1\ 10^{13}$	$1\ 10^6$
Ag-103	$1\ 10^1$	$1\ 10^6$	$1\ 10^{13}$	$1\ 10^7$
Ag-104	$1\ 10^1$	$1\ 10^6$	$1\ 10^{12}$	$1\ 10^7$
Ag-104m	$1\ 10^1$	$1\ 10^6$	$1\ 10^{13}$	$1\ 10^7$
Ag-105	$1\ 10^2$	$1\ 10^6$	$1\ 10^{11}$	$1\ 10^7$
Ag-106	$1\ 10^1$	$1\ 10^6$	$1\ 10^{13}$	$1\ 10^7$
Ag-106m	$1\ 10^1$	$1\ 10^6$	$1\ 10^{10}$	$1\ 10^7$
Ag-108m +	$1\ 10^1$	$1\ 10^6$	$1\ 10^{10}$	$1\ 10^7$
Ag-110m	$1\ 10^1$	$1\ 10^6$	$1\ 10^{10}$	$1\ 10^7$
Ag-111	$1\ 10^3$	$1\ 10^6$	$1\ 10^{11}$	$1\ 10^7$
Ag-112	$1\ 10^1$	$1\ 10^5$	$1\ 10^{12}$	$1\ 10^6$
Ag-115	$1\ 10^1$	$1\ 10^5$	$1\ 10^{13}$	$1\ 10^6$
Cadmium				
Cd-104	$1\ 10^2$	$1\ 10^7$	$1\ 10^{13}$	$1\ 10^8$
Cd-107	$1\ 10^3$	$1\ 10^7$	$1\ 10^{13}$	$1\ 10^8$
Cd-109	$1\ 10^4$	$1\ 10^6$	$1\ 10^{10}$	$1\ 10^7$
Cd-113	$1\ 10^3$	$1\ 10^6$	$1\ 10^9$	$1\ 10^7$
Cd-113m	$1\ 10^3$	$1\ 10^6$	$1\ 10^9$	$1\ 10^7$
Cd-115	$1\ 10^2$	$1\ 10^6$	$1\ 10^{11}$	$1\ 10^7$
Cd-115m	$1\ 10^3$	$1\ 10^6$	$1\ 10^{10}$	$1\ 10^7$
Cd-117	$1\ 10^1$	$1\ 10^6$	$1\ 10^{12}$	$1\ 10^7$
Cd-117m	$1\ 10^1$	$1\ 10^6$	$1\ 10^{12}$	$1\ 10^7$
Indium				
In-109	$1\ 10^1$	$1\ 10^6$	$1\ 10^{12}$	$1\ 10^7$
In-110 (4.9 hours)	$1\ 10^1$	$1\ 10^6$	$1\ 10^{12}$	$1\ 10^7$
In-110 (69.1 min)	$1\ 10^1$	$1\ 10^5$	$1\ 10^{13}$	$1\ 10^6$

1	2	3	4	5
Radionuclide name, symbol, isotope	**Concentration for notification. Regulation 6 and Schedule 1 (Bq/g)**	**Quantity for notification. Regulation 6 and Schedule 1 (Bq)**	**Quantity for notification of occurrences. Regulation 30(1) (Bq)**	**Quantity for notification of occurrences. Regulation 30(3) (Bq)**
In-111	$1\ 10^2$	$1\ 10^6$	$1\ 10^{11}$	$1\ 10^7$
In-112	$1\ 10^2$	$1\ 10^6$	$1\ 10^{14}$	$1\ 10^7$
In-113m	$1\ 10^2$	$1\ 10^6$	$1\ 10^{13}$	$1\ 10^7$
In-114	$1\ 10^3$	$1\ 10^5$	$1\ 10^{15}$	$1\ 10^6$
In-114m	$1\ 10^2$	$1\ 10^6$	$1\ 10^{10}$	$1\ 10^7$
In-115	$1\ 10^3$	$1\ 10^5$	$1\ 10^8$	$1\ 10^6$
In-115m	$1\ 10^2$	$1\ 10^6$	$1\ 10^{13}$	$1\ 10^7$
In-116m	$1\ 10^1$	$1\ 10^5$	$1\ 10^{13}$	$1\ 10^6$
In-117	$1\ 10^1$	$1\ 10^6$	$1\ 10^{13}$	$1\ 10^7$
In-117m	$1\ 10^2$	$1\ 10^6$	$1\ 10^{13}$	$1\ 10^7$
In-119m	$1\ 10^2$	$1\ 10^5$	$1\ 10^{14}$	$1\ 10^6$
Tin				
Sn-110	$1\ 10^2$	$1\ 10^7$	$1\ 10^{12}$	$1\ 10^8$
Sn-111	$1\ 10^2$	$1\ 10^6$	$1\ 10^{13}$	$1\ 10^7$
Sn-113	$1\ 10^3$	$1\ 10^7$	$1\ 10^{11}$	$1\ 10^8$
Sn-117m	$1\ 10^2$	$1\ 10^6$	$1\ 10^{11}$	$1\ 10^7$
Sn-119m	$1\ 10^3$	$1\ 10^7$	$1\ 10^{11}$	$1\ 10^8$
Sn-121	$1\ 10^5$	$1\ 10^7$	$1\ 10^{12}$	$1\ 10^8$
Sn-121m+	$1\ 10^3$	$1\ 10^7$	$1\ 10^{10}$	$1\ 10^8$
Sn-123	$1\ 10^3$	$1\ 10^6$	$1\ 10^{10}$	$1\ 10^7$
Sn-123m	$1\ 10^?$	$1\ 10^6$	$1\ 10^{14}$	$1\ 10^7$
Sn-125	$1\ 10^2$	$1\ 10^5$	$1\ 10^{10}$	$1\ 10^6$
Sn-126+	$1\ 10^1$	$1\ 10^5$	$1\ 10^{10}$	$1\ 10^6$
Sn-127	$1\ 10^1$	$1\ 10^6$	$1\ 10^{12}$	$1\ 10^7$
Sn-128	$1\ 10^1$	$1\ 10^6$	$1\ 10^{13}$	$1\ 10^7$
Antimony				
Sb-115	$1\ 10^1$	$1\ 10^6$	$1\ 10^{13}$	$1\ 10^7$
Sb-116	$1\ 10^1$	$1\ 10^6$	$1\ 10^{13}$	$1\ 10^7$
Sb-116m	$1\ 10^1$	$1\ 10^5$	$1\ 10^{12}$	$1\ 10^6$
Sb-117	$1\ 10^2$	$1\ 10^7$	$1\ 10^{13}$	$1\ 10^8$
Sb-118m	$1\ 10^1$	$1\ 10^6$	$1\ 10^{12}$	$1\ 10^7$
Sb-119	$1\ 10^3$	$1\ 10^7$	$1\ 10^{12}$	$1\ 10^8$
Sb-120 (5.76 days)	$1\ 10^1$	$1\ 10^6$	$1\ 10^{10}$	$1\ 10^7$
Sb-120 (15.89 min)	$1\ 10^2$	$1\ 10^6$	$1\ 10^{14}$	$1\ 10^7$

1	2	3	4	5
Radionuclide name, symbol, isotope	Concentration for notification. Regulation 6 and Schedule 1 (Bq/g)	Quantity for notification. Regulation 6 and Schedule 1 (Bq)	Quantity for notification of occurrences. Regulation 30(1) (Bq)	Quantity for notification of occurrences. Regulation 30(3) (Bq)
Sb-122	$1\ 10^2$	$1\ 10^4$	$1\ 10^{11}$	$1\ 10^5$
Sb-124	$1\ 10^1$	$1\ 10^6$	$1\ 10^{10}$	$1\ 10^7$
Sb-124m	$1\ 10^2$	$1\ 10^6$	$1\ 10^{14}$	$1\ 10^7$
Sb-125	$1\ 10^2$	$1\ 10^6$	$1\ 10^{10}$	$1\ 10^7$
Sb-126	$1\ 10^1$	$1\ 10^5$	$1\ 10^{10}$	$1\ 10^6$
Sb-126m	$1\ 10^1$	$1\ 10^5$	$1\ 10^{13}$	$1\ 10^6$
Sb-127	$1\ 10^1$	$1\ 10^6$	$1\ 10^{11}$	$1\ 10^7$
Sb-128 (9.01 hours)	$1\ 10^1$	$1\ 10^5$	$1\ 10^{11}$	$1\ 10^6$
Sb-128 (10.4 min)	$1\ 10^1$	$1\ 10^5$	$1\ 10^{13}$	$1\ 10^6$
Sb-129	$1\ 10^1$	$1\ 10^6$	$1\ 10^{12}$	$1\ 10^7$
Sb-130	$1\ 10^1$	$1\ 10^5$	$1\ 10^{13}$	$1\ 10^6$
Sb-131	$1\ 10^1$	$1\ 10^6$	$1\ 10^{13}$	$1\ 10^7$
Tellurium				
Te-116	$1\ 10^2$	$1\ 10^7$	$1\ 10^{13}$	$1\ 10^8$
Te-121	$1\ 10^1$	$1\ 10^6$	$1\ 10^{11}$	$1\ 10^7$
Te-121m	$1\ 10^2$	$1\ 10^6$	$1\ 10^{10}$	$1\ 10^7$
Te-123	$1\ 10^3$	$1\ 10^6$	$1\ 10^{10}$	$1\ 10^7$
Te-123m	$1\ 10^2$	$1\ 10^7$	$1\ 10^{10}$	$1\ 10^8$
Te-125m	$1\ 10^3$	$1\ 10^7$	$1\ 10^{10}$	$1\ 10^8$
Te-127	$1\ 10^3$	$1\ 10^6$	$1\ 10^{12}$	$1\ 10^7$
Te-127m	$1\ 10^3$	$1\ 10^7$	$1\ 10^{10}$	$1\ 10^8$
Te-129	$1\ 10^2$	$1\ 10^6$	$1\ 10^{14}$	$1\ 10^7$
Te-129m	$1\ 10^3$	$1\ 10^6$	$1\ 10^{10}$	$1\ 10^7$
Te-131	$1\ 10^2$	$1\ 10^5$	$1\ 10^{14}$	$1\ 10^6$
Te-131m	$1\ 10^1$	$1\ 10^6$	$1\ 10^{11}$	$1\ 10^7$
Te-132	$1\ 10^2$	$1\ 10^7$	$1\ 10^{11}$	$1\ 10^8$
Te-133	$1\ 10^1$	$1\ 10^5$	$1\ 10^{14}$	$1\ 10^6$
Te-133m	$1\ 10^1$	$1\ 10^5$	$1\ 10^{13}$	$1\ 10^6$
Te-134	$1\ 10^1$	$1\ 10^6$	$1\ 10^{13}$	$1\ 10^7$
Iodine				
I-120	$1\ 10^1$	$1\ 10^5$	$1\ 10^{12}$	$1\ 10^6$
I-120m	$1\ 10^1$	$1\ 10^5$	$1\ 10^{12}$	$1\ 10^6$
I-121	$1\ 10^2$	$1\ 10^6$	$1\ 10^{13}$	$1\ 10^7$
I-123	$1\ 10^2$	$1\ 10^7$	$1\ 10^{12}$	$1\ 10^8$

1	2	3	4	5
Radionuclide name, symbol, isotope	**Concentration for notification. Regulation 6 and Schedule 1 (Bq/g)**	**Quantity for notification. Regulation 6 and Schedule 1 (Bq)**	**Quantity for notification of occurrences. Regulation 30(1) (Bq)**	**Quantity for notification of occurrences. Regulation 30(3) (Bq)**
I-124	$1\ 10^1$	$1\ 10^6$	$1\ 10^{10}$	$1\ 10^7$
I-125	$1\ 10^3$	$1\ 10^6$	$1\ 10^{10}$	$1\ 10^7$
I-126	$1\ 10^2$	$1\ 10^6$	$1\ 10^{10}$	$1\ 10^7$
I-128	$1\ 10^2$	$1\ 10^5$	$1\ 10^{14}$	$1\ 10^6$
I-129	$1\ 10^2$	$1\ 10^5$	$1\ 10^9$	$1\ 10^6$
I-130	$1\ 10^1$	$1\ 10^6$	$1\ 10^{11}$	$1\ 10^7$
I-131	$1\ 10^2$	$1\ 10^6$	$1\ 10^{10}$	$1\ 10^7$
I-132	$1\ 10^1$	$1\ 10^5$	$1\ 10^{12}$	$1\ 10^6$
I-132m	$1\ 10^2$	$1\ 10^6$	$1\ 10^{13}$	$1\ 10^7$
I-133	$1\ 10^1$	$1\ 10^6$	$1\ 10^{11}$	$1\ 10^7$
I-134	$1\ 10^1$	$1\ 10^5$	$1\ 10^{13}$	$1\ 10^6$
I-135	$1\ 10^1$	$1\ 10^6$	$1\ 10^{12}$	$1\ 10^7$
Xenon				
Xe-120	$1\ 10^2$	$1\ 10^9$	$1\ 10^{10}$	
Xe-121	$1\ 10^2$	$1\ 10^9$	$1\ 10^9$	
Xe-122 +	$1\ 10^2$	$1\ 10^9$	$1\ 10^{11}$	
Xe-123	$1\ 10^2$	$1\ 10^9$	$1\ 10^9$	
Xe-125	$1\ 10^3$	$1\ 10^9$	$1\ 10^{10}$	
Xe-127	$1\ 10^3$	$1\ 10^5$	$1\ 10^{10}$	
Xe-129m	$1\ 10^3$	$1\ 10^4$	$1\ 10^{11}$	
Xe-131m	$1\ 10^4$	$1\ 10^4$	$1\ 10^{11}$	
Xe-133	$1\ 10^3$	$1\ 10^4$	$1\ 10^{11}$	
Xe-133m	$1\ 10^3$	$1\ 10^4$	$1\ 10^{11}$	
Xe-135	$1\ 10^3$	$1\ 10^{10}$	$1\ 10^{10}$	
Xe-135m	$1\ 10^2$	$1\ 10^9$	$1\ 10^{10}$	
Xe-138	$1\ 10^2$	$1\ 10^9$	$1\ 10^9$	
Caesium				
Cs-125	$1\ 10^1$	$1\ 10^4$	$1\ 10^{13}$	$1\ 10^5$
Cs-127	$1\ 10^2$	$1\ 10^5$	$1\ 10^{12}$	$1\ 10^6$
Cs-129	$1\ 10^2$	$1\ 10^5$	$1\ 10^{12}$	$1\ 10^6$
Cs-130	$1\ 10^2$	$1\ 10^6$	$1\ 10^{14}$	$1\ 10^7$
Cs-131	$1\ 10^3$	$1\ 10^6$	$1\ 10^{12}$	$1\ 10^7$
Cs-132	$1\ 10^1$	$1\ 10^5$	$1\ 10^{11}$	$1\ 10^6$
Cs-134	$1\ 10^1$	$1\ 10^4$	$1\ 10^{10}$	$1\ 10^5$

1	2	3	4	5
Radionuclide name, symbol, isotope	Concentration for notification. Regulation 6 and Schedule 1 (Bq/g)	Quantity for notification. Regulation 6 and Schedule 1 (Bq)	Quantity for notification of occurrences. Regulation 30(1) (Bq)	Quantity for notification of occurrences. Regulation 30(3) (Bq)
Cs-134m	$1\ 10^3$	$1\ 10^5$	$1\ 10^{14}$	$1\ 10^6$
Cs-135	$1\ 10^4$	$1\ 10^7$	$1\ 10^{11}$	$1\ 10^8$
Cs-135m	$1\ 10^1$	$1\ 10^6$	$1\ 10^{13}$	$1\ 10^7$
Cs-136	$1\ 10^1$	$1\ 10^5$	$1\ 10^{10}$	$1\ 10^6$
Cs-137+	$1\ 10^1$	$1\ 10^4$	$1\ 10^{10}$	$1\ 10^5$
Cs-138	$1\ 10^1$	$1\ 10^4$	$1\ 10^{13}$	$1\ 10^5$
Barium				
Ba-126	$1\ 10^2$	$1\ 10^7$	$1\ 10^{13}$	$1\ 10^8$
Ba-128	$1\ 10^2$	$1\ 10^7$	$1\ 10^{11}$	$1\ 10^8$
Ba-131	$1\ 10^2$	$1\ 10^6$	$1\ 10^{11}$	$1\ 10^7$
Ba-131m	$1\ 10^2$	$1\ 10^7$	$1\ 10^{15}$	$1\ 10^8$
Ba-133	$1\ 10^2$	$1\ 10^6$	$1\ 10^{11}$	$1\ 10^7$
Ba-133m	$1\ 10^2$	$1\ 10^6$	$1\ 10^{12}$	$1\ 10^7$
Ba-135m	$1\ 10^2$	$1\ 10^6$	$1\ 10^{12}$	$1\ 10^7$
Ba-137m	$1\ 10^1$	$1\ 10^6$	$1\ 10^{15}$	$1\ 10^7$
Ba-139	$1\ 10^2$	$1\ 10^5$	$1\ 10^{13}$	$1\ 10^6$
Ba-140+	$1\ 10^1$	$1\ 10^5$	$1\ 10^{11}$	$1\ 10^6$
Ba-141	$1\ 10^1$	$1\ 10^5$	$1\ 10^{13}$	$1\ 10^6$
Ba-142	$1\ 10^1$	$1\ 10^6$	$1\ 10^{14}$	$1\ 10^7$
Lanthanum				
La-131	$1\ 10^1$	$1\ 10^6$	$1\ 10^{13}$	$1\ 10^7$
La-132	$1\ 10^1$	$1\ 10^6$	$1\ 10^{12}$	$1\ 10^7$
La-135	$1\ 10^3$	$1\ 10^7$	$1\ 10^{13}$	$1\ 10^8$
La-137	$1\ 10^3$	$1\ 10^7$	$1\ 10^{10}$	$1\ 10^8$
La-138	$1\ 10^1$	$1\ 10^6$	$1\ 10^9$	$1\ 10^7$
La-140	$1\ 10^1$	$1\ 10^5$	$1\ 10^{11}$	$1\ 10^6$
La-141	$1\ 10^2$	$1\ 10^5$	$1\ 10^{13}$	$1\ 10^6$
La-142	$1\ 10^1$	$1\ 10^5$	$1\ 10^{12}$	$1\ 10^6$
La-143	$1\ 10^2$	$1\ 10^5$	$1\ 10^{14}$	$1\ 10^6$
Cerium				
Ce-134	$1\ 10^3$	$1\ 10^7$	$1\ 10^{11}$	$1\ 10^8$
Ce-135	$1\ 10^1$	$1\ 10^6$	$1\ 10^{11}$	$1\ 10^7$
Ce-137	$1\ 10^3$	$1\ 10^7$	$1\ 10^{13}$	$1\ 10^8$

1	2	3	4	5
Radionuclide name, symbol, isotope	Concentration for notification. Regulation 6 and Schedule 1 (Bq/g)	Quantity for notification. Regulation 6 and Schedule 1 (Bq)	Quantity for notification of occurrences. Regulation 30(1) (Bq)	Quantity for notification of occurrences. Regulation 30(3) (Bq)
Ce-137m	$1\ 10^3$	$1\ 10^6$	$1\ 10^{11}$	$1\ 10^7$
Ce-139	$1\ 10^2$	$1\ 10^6$	$1\ 10^{11}$	$1\ 10^7$
Ce-141	$1\ 10^2$	$1\ 10^7$	$1\ 10^{10}$	$1\ 10^8$
Ce-143	$1\ 10^2$	$1\ 10^6$	$1\ 10^{11}$	$1\ 10^7$
Ce-144+	$1\ 10^2$	$1\ 10^5$	$1\ 10^9$	$1\ 10^6$
Praseodymium				
Pr-136	$1\ 10^1$	$1\ 10^5$	$1\ 10^{13}$	$1\ 10^6$
Pr-137	$1\ 10^2$	$1\ 10^6$	$1\ 10^{13}$	$1\ 10^7$
Pr-138m	$1\ 10^1$	$1\ 10^6$	$1\ 10^{12}$	$1\ 10^7$
Pr-139	$1\ 10^2$	$1\ 10^7$	$1\ 10^{13}$	$1\ 10^8$
Pr-142	$1\ 10^2$	$1\ 10^5$	$1\ 10^{12}$	$1\ 10^6$
Pr-142m	$1\ 10^7$	$1\ 10^9$	$1\ 10^{15}$	$1\ 10^{10}$
Pr-143	$1\ 10^4$	$1\ 10^6$	$1\ 10^{11}$	$1\ 10^7$
Pr-144	$1\ 10^2$	$1\ 10^5$	$1\ 10^{14}$	$1\ 10^6$
Pr-145	$1\ 10^3$	$1\ 10^5$	$1\ 10^{12}$	$1\ 10^6$
Pr-147	$1\ 10^1$	$1\ 10^5$	$1\ 10^{14}$	$1\ 10^6$
Neodymium				
Nd-136	$1\ 10^2$	$1\ 10^6$	$1\ 10^{13}$	$1\ 10^7$
Nd-138	$1\ 10^3$	$1\ 10^7$	$1\ 10^{12}$	$1\ 10^8$
Nd-139	$1\ 10^2$	$1\ 10^6$	$1\ 10^{14}$	$1\ 10^7$
Nd-139m	$1\ 10^1$	$1\ 10^6$	$1\ 10^{12}$	$1\ 10^7$
Nd-141	$1\ 10^2$	$1\ 10^7$	$1\ 10^{14}$	$1\ 10^8$
Nd-147	$1\ 10^2$	$1\ 10^6$	$1\ 10^{11}$	$1\ 10^7$
Nd-149	$1\ 10^2$	$1\ 10^6$	$1\ 10^{13}$	$1\ 10^7$
Nd-151	$1\ 10^1$	$1\ 10^5$	$1\ 10^{14}$	$1\ 10^6$
Promethium				
Pm-141	$1\ 10^1$	$1\ 10^5$	$1\ 10^{13}$	$1\ 10^6$
Pm-143	$1\ 10^2$	$1\ 10^6$	$1\ 10^{11}$	$1\ 10^7$
Pm-144	$1\ 10^1$	$1\ 10^6$	$1\ 10^{10}$	$1\ 10^7$
Pm-145	$1\ 10^3$	$1\ 10^7$	$1\ 10^{10}$	$1\ 10^8$
Pm-146	$1\ 10^1$	$1\ 10^6$	$1\ 10^{10}$	$1\ 10^7$
Pm-147	$1\ 10^4$	$1\ 10^7$	$1\ 10^{10}$	$1\ 10^8$
Pm-148	$1\ 10^1$	$1\ 10^5$	$1\ 10^{11}$	$1\ 10^6$

1	2	3	4	5
Radionuclide name, symbol, isotope	**Concentration for notification. Regulation 6 and Schedule 1 (Bq/g)**	**Quantity for notification. Regulation 6 and Schedule 1 (Bq)**	**Quantity for notification of occurrences. Regulation 30(1) (Bq)**	**Quantity for notification of occurrences. Regulation 30(3) (Bq)**
Pm-148m+	$1\ 10^{1}$	$1\ 10^{6}$	$1\ 10^{10}$	$1\ 10^{7}$
Pm-149	$1\ 10^{3}$	$1\ 10^{6}$	$1\ 10^{11}$	$1\ 10^{7}$
Pm-150	$1\ 10^{1}$	$1\ 10^{5}$	$1\ 10^{12}$	$1\ 10^{6}$
Pm-151	$1\ 10^{2}$	$1\ 10^{6}$	$1\ 10^{11}$	$1\ 10^{7}$
Samarium				
Sm-141	$1\ 10^{1}$	$1\ 10^{5}$	$1\ 10^{13}$	$1\ 10^{6}$
Sm-141m	$1\ 10^{1}$	$1\ 10^{6}$	$1\ 10^{13}$	$1\ 10^{7}$
Sm-142	$1\ 10^{2}$	$1\ 10^{7}$	$1\ 10^{13}$	$1\ 10^{8}$
Sm-145	$1\ 10^{2}$	$1\ 10^{7}$	$1\ 10^{11}$	$1\ 10^{8}$
Sm-146	$1\ 10^{1}$	$1\ 10^{5}$	$1\ 10^{7}$	$1\ 10^{6}$
Sm-147	$1\ 10^{1}$	$1\ 10^{4}$	$1\ 10^{7}$	$1\ 10^{5}$
Sm-151	$1\ 10^{4}$	$1\ 10^{8}$	$1\ 10^{10}$	$1\ 10^{9}$
Sm-153	$1\ 10^{2}$	$1\ 10^{6}$	$1\ 10^{11}$	$1\ 10^{7}$
Sm-155	$1\ 10^{2}$	$1\ 10^{6}$	$1\ 10^{14}$	$1\ 10^{7}$
Sm-156	$1\ 10^{2}$	$1\ 10^{6}$	$1\ 10^{12}$	$1\ 10^{7}$
Europium				
Eu-145	$1\ 10^{1}$	$1\ 10^{6}$	$1\ 10^{11}$	$1\ 10^{7}$
Eu-146	$1\ 10^{1}$	$1\ 10^{6}$	$1\ 10^{11}$	$1\ 10^{7}$
Eu-147	$1\ 10^{2}$	$1\ 10^{6}$	$1\ 10^{11}$	$1\ 10^{7}$
Eu-148	$1\ 10^{1}$	$1\ 10^{6}$	$1\ 10^{10}$	$1\ 10^{7}$
Eu-149	$1\ 10^{2}$	$1\ 10^{7}$	$1\ 10^{11}$	$1\ 10^{8}$
Eu-150 (34.2 years)	$1\ 10^{1}$	$1\ 10^{6}$	$1\ 10^{9}$	$1\ 10^{7}$
Eu-150 (12.6 hours)	$1\ 10^{3}$	$1\ 10^{6}$	$1\ 10^{12}$	$1\ 10^{7}$
Eu-152	$1\ 10^{1}$	$1\ 10^{6}$	$1\ 10^{9}$	$1\ 10^{7}$
Eu-152m	$1\ 10^{2}$	$1\ 10^{6}$	$1\ 10^{12}$	$1\ 10^{7}$
Eu-154	$1\ 10^{1}$	$1\ 10^{6}$	$1\ 10^{9}$	$1\ 10^{7}$
Eu-155	$1\ 10^{2}$	$1\ 10^{7}$	$1\ 10^{10}$	$1\ 10^{8}$
Eu-156	$1\ 10^{1}$	$1\ 10^{6}$	$1\ 10^{10}$	$1\ 10^{7}$
Eu-157	$1\ 10^{2}$	$1\ 10^{6}$	$1\ 10^{12}$	$1\ 10^{7}$
Eu-158	$1\ 10^{1}$	$1\ 10^{5}$	$1\ 10^{13}$	$1\ 10^{6}$

1	2	3	4	5
Radionuclide name, symbol, isotope	Concentration for notification. Regulation 6 and Schedule 1 (Bq/g)	Quantity for notification. Regulation 6 and Schedule 1 (Bq)	Quantity for notification of occurrences. Regulation 30(1) (Bq)	Quantity for notification of occurrences. Regulation 30(3) (Bq)
Gadolinium				
Gd-145	$1\ 10^1$	$1\ 10^5$	$1\ 10^{13}$	$1\ 10^6$
Gd-146 +	$1\ 10^1$	$1\ 10^6$	$1\ 10^{10}$	$1\ 10^7$
Gd-147	$1\ 10^1$	$1\ 10^6$	$1\ 10^{11}$	$1\ 10^7$
Gd-148	$1\ 10^1$	$1\ 10^4$	$1\ 10^6$	$1\ 10^5$
Gd-149	$1\ 10^2$	$1\ 10^6$	$1\ 10^{11}$	$1\ 10^7$
Gd-151	$1\ 10^2$	$1\ 10^7$	$1\ 10^{11}$	$1\ 10^8$
Gd-152	$1\ 10^1$	$1\ 10^4$	$1\ 10^6$	$1\ 10^5$
Gd-153	$1\ 10^2$	$1\ 10^7$	$1\ 10^{10}$	$1\ 10^8$
Gd-159	$1\ 10^3$	$1\ 10^6$	$1\ 10^{12}$	$1\ 10^7$
Terbium				
Tb-147	$1\ 10^1$	$1\ 10^6$	$1\ 10^{12}$	$1\ 10^7$
Tb-149	$1\ 10^1$	$1\ 10^6$	$1\ 10^{11}$	$1\ 10^7$
Tb-150	$1\ 10^1$	$1\ 10^6$	$1\ 10^{12}$	$1\ 10^7$
Tb-151	$1\ 10^1$	$1\ 10^6$	$1\ 10^{12}$	$1\ 10^7$
Tb-153	$1\ 10^2$	$1\ 10^7$	$1\ 10^{12}$	$1\ 10^8$
Tb-154	$1\ 10^1$	$1\ 10^6$	$1\ 10^{11}$	$1\ 10^7$
Tb-155	$1\ 10^2$	$1\ 10^7$	$1\ 10^{11}$	$1\ 10^8$
Tb-156	$1\ 10^1$	$1\ 10^6$	$1\ 10^{11}$	$1\ 10^7$
Tb-156m (24.4 hours)	$1\ 10^3$	$1\ 10^7$	$1\ 10^{12}$	$1\ 10^8$
Tb-156m (5 hours)	$1\ 10^4$	$1\ 10^7$	$1\ 10^{13}$	$1\ 10^8$
Tb-157	$1\ 10^4$	$1\ 10^7$	$1\ 10^{11}$	$1\ 10^8$
Tb-158	$1\ 10^1$	$1\ 10^6$	$1\ 10^9$	$1\ 10^7$
Tb-160	$1\ 10^1$	$1\ 10^6$	$1\ 10^{10}$	$1\ 10^7$
Tb-161	$1\ 10^3$	$1\ 10^6$	$1\ 10^{11}$	$1\ 10^7$
Dysprosium				
Dy-155	$1\ 10^1$	$1\ 10^6$	$1\ 10^{12}$	$1\ 10^7$
Dy-157	$1\ 10^2$	$1\ 10^6$	$1\ 10^{12}$	$1\ 10^7$
Dy-159	$1\ 10^3$	$1\ 10^7$	$1\ 10^{11}$	$1\ 10^8$
Dy-165	$1\ 10^3$	$1\ 10^6$	$1\ 10^{13}$	$1\ 10^7$
Dy-166	$1\ 10^3$	$1\ 10^6$	$1\ 10^{11}$	$1\ 10^7$

1	2	3	4	5
Radionuclide name, symbol, isotope	Concentration for notification. Regulation 6 and Schedule 1 (Bq/g)	Quantity for notification. Regulation 6 and Schedule 1 (Bq)	Quantity for notification of occurrences. Regulation 30(1) (Bq)	Quantity for notification of occurrences. Regulation 30(3) (Bq)
Holmium				
Ho-155	$1\ 10^2$	$1\ 10^6$	$1\ 10^{13}$	$1\ 10^7$
Ho-157	$1\ 10^2$	$1\ 10^6$	$1\ 10^{14}$	$1\ 10^7$
Ho-159	$1\ 10^2$	$1\ 10^6$	$1\ 10^{14}$	$1\ 10^7$
Ho-161	$1\ 10^2$	$1\ 10^7$	$1\ 10^{14}$	$1\ 10^8$
Ho-162	$1\ 10^2$	$1\ 10^7$	$1\ 10^{14}$	$1\ 10^8$
Ho-162m	$1\ 10^1$	$1\ 10^6$	$1\ 10^{13}$	$1\ 10^7$
Ho-164	$1\ 10^3$	$1\ 10^6$	$1\ 10^{14}$	$1\ 10^7$
Ho-164m	$1\ 10^3$	$1\ 10^7$	$1\ 10^{14}$	$1\ 10^8$
Ho-166	$1\ 10^3$	$1\ 10^5$	$1\ 10^{11}$	$1\ 10^6$
Ho-166m	$1\ 10^1$	$1\ 10^6$	$1\ 10^9$	$1\ 10^7$
Ho-167	$1\ 10^2$	$1\ 10^6$	$1\ 10^{13}$	$1\ 10^7$
Erbium				
Er-161	$1\ 10^1$	$1\ 10^6$	$1\ 10^{12}$	$1\ 10^7$
Er-165	$1\ 10^3$	$1\ 10^7$	$1\ 10^{13}$	$1\ 10^8$
Er-169	$1\ 10^4$	$1\ 10^7$	$1\ 10^{11}$	$1\ 10^8$
Er-171	$1\ 10^2$	$1\ 10^6$	$1\ 10^{12}$	$1\ 10^7$
Er-172	$1\ 10^2$	$1\ 10^6$	$1\ 10^{11}$	$1\ 10^7$
Thulium				
Tm-162	$1\ 10^1$	$1\ 10^6$	$1\ 10^{13}$	$1\ 10^7$
Tm-166	$1\ 10^1$	$1\ 10^6$	$1\ 10^{12}$	$1\ 10^7$
Tm-167	$1\ 10^2$	$1\ 10^6$	$1\ 10^{11}$	$1\ 10^7$
Tm-170	$1\ 10^3$	$1\ 10^6$	$1\ 10^{10}$	$1\ 10^7$
Tm-171	$1\ 10^4$	$1\ 10^8$	$1\ 10^{11}$	$1\ 10^9$
Tm-172	$1\ 10^2$	$1\ 10^6$	$1\ 10^{11}$	$1\ 10^7$
Tm-173	$1\ 10^2$	$1\ 10^6$	$1\ 10^{12}$	$1\ 10^7$
Tm-175	$1\ 10^1$	$1\ 10^6$	$1\ 10^{13}$	$1\ 10^7$
Ytterbium				
Yb-162	$1\ 10^2$	$1\ 10^7$	$1\ 10^{14}$	$1\ 10^8$
Yb-166	$1\ 10^2$	$1\ 10^7$	$1\ 10^{11}$	$1\ 10^8$
Yb-167	$1\ 10^2$	$1\ 10^6$	$1\ 10^{14}$	$1\ 10^7$
Yb-169	$1\ 10^2$	$1\ 10^7$	$1\ 10^{10}$	$1\ 10^8$
Yb-175	$1\ 10^3$	$1\ 10^7$	$1\ 10^{11}$	$1\ 10^8$

1	2	3	4	5
Radionuclide name, symbol, isotope	Concentration for notification. Regulation 6 and Schedule 1 (Bq/g)	Quantity for notification. Regulation 6 and Schedule 1 (Bq)	Quantity for notification of occurrences. Regulation 30(1) (Bq)	Quantity for notification of occurrences. Regulation 30(3) (Bq)
Yb-177	$1\ 10^2$	$1\ 10^6$	$1\ 10^{13}$	$1\ 10^7$
Yb-178	$1\ 10^3$	$1\ 10^6$	$1\ 10^{13}$	$1\ 10^7$
Lutetium				
Lu-169	$1\ 10^1$	$1\ 10^6$	$1\ 10^{11}$	$1\ 10^7$
Lu-170	$1\ 10^1$	$1\ 10^6$	$1\ 10^{11}$	$1\ 10^7$
Lu-171	$1\ 10^1$	$1\ 10^6$	$1\ 10^{11}$	$1\ 10^7$
Lu-172	$1\ 10^1$	$1\ 10^6$	$1\ 10^{10}$	$1\ 10^7$
Lu-173	$1\ 10^2$	$1\ 10^7$	$1\ 10^{11}$	$1\ 10^8$
Lu-174	$1\ 10^2$	$1\ 10^7$	$1\ 10^{10}$	$1\ 10^8$
Lu-174m	$1\ 10^2$	$1\ 10^7$	$1\ 10^{10}$	$1\ 10^8$
Lu-176	$1\ 10^2$	$1\ 10^6$	$1\ 10^9$	$1\ 10^7$
Lu-176m	$1\ 10^3$	$1\ 10^6$	$1\ 10^{13}$	$1\ 10^7$
Lu-177	$1\ 10^3$	$1\ 10^7$	$1\ 10^{11}$	$1\ 10^8$
Lu-177m	$1\ 10^1$	$1\ 10^6$	$1\ 10^{10}$	$1\ 10^7$
Lu-178	$1\ 10^2$	$1\ 10^5$	$1\ 10^{14}$	$1\ 10^6$
Lu-178m	$1\ 10^1$	$1\ 10^5$	$1\ 10^{13}$	$1\ 10^6$
Lu-179	$1\ 10^3$	$1\ 10^6$	$1\ 10^{13}$	$1\ 10^7$
Hafnium				
Hf-170	$1\ 10^2$	$1\ 10^6$	$1\ 10^{12}$	$1\ 10^7$
Hf-172+	$1\ 10^1$	$1\ 10^6$	$1\ 10^9$	$1\ 10^7$
Hf-173	$1\ 10^2$	$1\ 10^6$	$1\ 10^{12}$	$1\ 10^7$
Hf-175	$1\ 10^2$	$1\ 10^6$	$1\ 10^{11}$	$1\ 10^7$
Hf-177m	$1\ 10^1$	$1\ 10^5$	$1\ 10^{13}$	$1\ 10^6$
Hf-178m	$1\ 10^1$	$1\ 10^6$	$1\ 10^8$	$1\ 10^7$
Hf-179m	$1\ 10^1$	$1\ 10^6$	$1\ 10^{10}$	$1\ 10^7$
Hf-180m	$1\ 10^1$	$1\ 10^6$	$1\ 10^{12}$	$1\ 10^7$
Hf-181	$1\ 10^1$	$1\ 10^6$	$1\ 10^{10}$	$1\ 10^7$
Hf-182	$1\ 10^2$	$1\ 10^6$	$1\ 10^8$	$1\ 10^7$
Hf-182m	$1\ 10^1$	$1\ 10^6$	$1\ 10^{13}$	$1\ 10^7$
Hf-183	$1\ 10^1$	$1\ 10^6$	$1\ 10^{13}$	$1\ 10^7$
Hf-184	$1\ 10^2$	$1\ 10^6$	$1\ 10^{12}$	$1\ 10^7$
Tantalum				
Ta-172	$1\ 10^1$	$1\ 10^6$	$1\ 10^{13}$	$1\ 10^7$

1	2	3	4	5
Radionuclide name, symbol, isotope	Concentration for notification. Regulation 6 and Schedule 1 (Bq/g)	Quantity for notification. Regulation 6 and Schedule 1 (Bq)	Quantity for notification of occurrences. Regulation 30(1) (Bq)	Quantity for notification of occurrences. Regulation 30(3) (Bq)
Ta-173	$1\ 10^1$	$1\ 10^6$	$1\ 10^{12}$	$1\ 10^7$
Ta-174	$1\ 10^1$	$1\ 10^6$	$1\ 10^{13}$	$1\ 10^7$
Ta-175	$1\ 10^1$	$1\ 10^6$	$1\ 10^{10}$	$1\ 10^7$
Ta-176	$1\ 10^1$	$1\ 10^6$	$1\ 10^{12}$	$1\ 10^7$
Ta-177	$1\ 10^2$	$1\ 10^7$	$1\ 10^{12}$	$1\ 10^8$
Ta-178	$1\ 10^1$	$1\ 10^6$	$1\ 10^{13}$	$1\ 10^7$
Ta-179	$1\ 10^3$	$1\ 10^7$	$1\ 10^{11}$	$1\ 10^8$
Ta-180	$1\ 10^1$	$1\ 10^6$	$1\ 10^{10}$	$1\ 10^7$
Ta-180m	$1\ 10^3$	$1\ 10^7$	$1\ 10^{13}$	$1\ 10^8$
Ta-182	$1\ 10^1$	$1\ 10^4$	$1\ 10^{10}$	$1\ 10^5$
Ta-182m	$1\ 10^2$	$1\ 10^6$	$1\ 10^{14}$	$1\ 10^7$
Ta-183	$1\ 10^2$	$1\ 10^6$	$1\ 10^{11}$	$1\ 10^7$
Ta-184	$1\ 10^1$	$1\ 10^6$	$1\ 10^{12}$	$1\ 10^7$
Ta-185	$1\ 10^2$	$1\ 10^5$	$1\ 10^{13}$	$1\ 10^6$
Ta-186	$1\ 10^1$	$1\ 10^5$	$1\ 10^{13}$	$1\ 10^6$
Tungsten				
W-176	$1\ 10^2$	$1\ 10^6$	$1\ 10^{13}$	$1\ 10^7$
W-177	$1\ 10^1$	$1\ 10^6$	$1\ 10^{13}$	$1\ 10^7$
W-178 +	$1\ 10^1$	$1\ 10^6$	$1\ 10^{12}$	$1\ 10^7$
W-179	$1\ 10^2$	$1\ 10^7$	$1\ 10^{14}$	$1\ 10^8$
W-181	$1\ 10^3$	$1\ 10^7$	$1\ 10^{12}$	$1\ 10^8$
W-185	$1\ 10^4$	$1\ 10^7$	$1\ 10^{11}$	$1\ 10^8$
W-187	$1\ 10^2$	$1\ 10^6$	$1\ 10^{12}$	$1\ 10^7$
W-188 +	$1\ 10^2$	$1\ 10^5$	$1\ 10^{11}$	$1\ 10^6$
Rhenium				
Re-177	$1\ 10^1$	$1\ 10^6$	$1\ 10^{14}$	$1\ 10^7$
Re-178	$1\ 10^1$	$1\ 10^6$	$1\ 10^{13}$	$1\ 10^7$
Re-181	$1\ 10^1$	$1\ 10^6$	$1\ 10^{11}$	$1\ 10^7$
Re-182 (64 hours)	$1\ 10^1$	$1\ 10^6$	$1\ 10^{11}$	$1\ 10^7$
Re-182 (12.7 hours)	$1\ 10^1$	$1\ 10^6$	$1\ 10^{12}$	$1\ 10^7$
Re-184	$1\ 10^1$	$1\ 10^6$	$1\ 10^{10}$	$1\ 10^7$
Re-184m	$1\ 10^2$	$1\ 10^6$	$1\ 10^{10}$	$1\ 10^7$
Re-186	$1\ 10^3$	$1\ 10^6$	$1\ 10^{11}$	$1\ 10^7$
Re-186m	$1\ 10^3$	$1\ 10^7$	$1\ 10^{10}$	$1\ 10^8$

1	2	3	4	5
Radionuclide name, symbol, isotope	Concentration for notification. Regulation 6 and Schedule 1 (Bq/g)	Quantity for notification. Regulation 6 and Schedule 1 (Bq)	Quantity for notification of occurrences. Regulation 30(1) (Bq)	Quantity for notification of occurrences. Regulation 30(3) (Bq)
Re-187	$1\ 10^6$	$1\ 10^9$	$1\ 10^{13}$	$1\ 10^{10}$
Re-188	$1\ 10^2$	$1\ 10^5$	$1\ 10^{12}$	$1\ 10^6$
Re-188m	$1\ 10^2$	$1\ 10^7$	$1\ 10^{14}$	$1\ 10^8$
Re-189+	$1\ 10^2$	$1\ 10^6$	$1\ 10^{12}$	$1\ 10^7$
Osmium				
Os-180	$1\ 10^2$	$1\ 10^7$	$1\ 10^{14}$	$1\ 10^8$
Os-181	$1\ 10^1$	$1\ 10^6$	$1\ 10^{13}$	$1\ 10^7$
Os-182	$1\ 10^2$	$1\ 10^6$	$1\ 10^{11}$	$1\ 10^7$
Os-185	$1\ 10^1$	$1\ 10^6$	$1\ 10^{11}$	$1\ 10^7$
Os-189m	$1\ 10^4$	$1\ 10^7$	$1\ 10^{14}$	$1\ 10^8$
Os-191	$1\ 10^2$	$1\ 10^7$	$1\ 10^{11}$	$1\ 10^8$
Os-191m	$1\ 10^3$	$1\ 10^7$	$1\ 10^{12}$	$1\ 10^8$
Os-193	$1\ 10^2$	$1\ 10^6$	$1\ 10^{11}$	$1\ 10^7$
Os-194+	$1\ 10^2$	$1\ 10^5$	$1\ 10^9$	$1\ 10^6$
Iridium				
Ir-182	$1\ 10^1$	$1\ 10^5$	$1\ 10^{13}$	$1\ 10^6$
Ir-184	$1\ 10^1$	$1\ 10^6$	$1\ 10^{12}$	$1\ 10^7$
Ir-185	$1\ 10^1$	$1\ 10^6$	$1\ 10^{12}$	$1\ 10^7$
Ir-186 (15.8 hours)	$1\ 10^1$	$1\ 10^6$	$1\ 10^{11}$	$1\ 10^7$
Ir-186 (1.75 hours)	$1\ 10^1$	$1\ 10^6$	$1\ 10^{13}$	$1\ 10^7$
Ir-187	$1\ 10^2$	$1\ 10^6$	$1\ 10^{12}$	$1\ 10^7$
Ir-188	$1\ 10^1$	$1\ 10^6$	$1\ 10^{11}$	$1\ 10^7$
Ir-189+	$1\ 10^2$	$1\ 10^7$	$1\ 10^{11}$	$1\ 10^8$
Ir-190	$1\ 10^1$	$1\ 10^6$	$1\ 10^{10}$	$1\ 10^7$
Ir-190m (3.1 hours)	$1\ 10^1$	$1\ 10^6$	$1\ 10^{13}$	$1\ 10^7$
Ir-190m (1.2 hours)	$1\ 10^4$	$1\ 10^7$	$1\ 10^{15}$	$1\ 10^8$
Ir-192	$1\ 10^1$	$1\ 10^4$	$1\ 10^{10}$	$1\ 10^5$
Ir-192m	$1\ 10^2$	$1\ 10^7$	$1\ 10^{10}$	$1\ 10^8$
Ir-193m	$1\ 10^4$	$1\ 10^7$	$1\ 10^{11}$	$1\ 10^8$
Ir-194	$1\ 10^2$	$1\ 10^5$	$1\ 10^{11}$	$1\ 10^6$
Ir-194m	$1\ 10^1$	$1\ 10^6$	$1\ 10^{10}$	$1\ 10^7$
Ir-195	$1\ 10^2$	$1\ 10^6$	$1\ 10^{13}$	$1\ 10^7$
Ir-195m	$1\ 10^2$	$1\ 10^6$	$1\ 10^{12}$	$1\ 10^7$

1	2	3	4	5
Radionuclide name, symbol, isotope	Concentration for notification. Regulation 6 and Schedule 1 (Bq/g)	Quantity for notification. Regulation 6 and Schedule 1 (Bq)	Quantity for notification of occurrences. Regulation 30(1) (Bq)	Quantity for notification of occurrences. Regulation 30(3) (Bq)
Platinum				
Pt-186	$1\ 10^1$	$1\ 10^6$	$1\ 10^{13}$	$1\ 10^7$
Pt-188 +	$1\ 10^1$	$1\ 10^6$	$1\ 10^{11}$	$1\ 10^7$
Pt-189	$1\ 10^2$	$1\ 10^6$	$1\ 10^{12}$	$1\ 10^7$
Pt-191	$1\ 10^2$	$1\ 10^6$	$1\ 10^{11}$	$1\ 10^7$
Pt-193	$1\ 10^4$	$1\ 10^7$	$1\ 10^{12}$	$1\ 10^8$
Pt-193m	$1\ 10^3$	$1\ 10^7$	$1\ 10^{12}$	$1\ 10^8$
Pt-195m	$1\ 10^2$	$1\ 10^6$	$1\ 10^{11}$	$1\ 10^7$
Pt-197	$1\ 10^3$	$1\ 10^6$	$1\ 10^{12}$	$1\ 10^7$
Pt-197m	$1\ 10^2$	$1\ 10^6$	$1\ 10^{14}$	$1\ 10^7$
Pt-199	$1\ 10^2$	$1\ 10^6$	$1\ 10^{14}$	$1\ 10^7$
Pt-200	$1\ 10^2$	$1\ 10^6$	$1\ 10^{12}$	$1\ 10^7$
Gold				
Au-193	$1\ 10^2$	$1\ 10^7$	$1\ 10^{12}$	$1\ 10^8$
Au-194	$1\ 10^1$	$1\ 10^6$	$1\ 10^{11}$	$1\ 10^7$
Au-195	$1\ 10^2$	$1\ 10^7$	$1\ 10^{11}$	$1\ 10^8$
Au-198	$1\ 10^2$	$1\ 10^6$	$1\ 10^{11}$	$1\ 10^7$
Au-198m	$1\ 10^1$	$1\ 10^6$	$1\ 10^{11}$	$1\ 10^7$
Au-199	$1\ 10^2$	$1\ 10^6$	$1\ 10^{11}$	$1\ 10^7$
Au-200	$1\ 10^2$	$1\ 10^5$	$1\ 10^{13}$	$1\ 10^6$
Au-200m	$1\ 10^1$	$1\ 10^6$	$1\ 10^{11}$	$1\ 10^7$
Au-201	$1\ 10^2$	$1\ 10^6$	$1\ 10^{14}$	$1\ 10^7$
Mercury				
Hg-193	$1\ 10^2$	$1\ 10^6$	$1\ 10^{13}$	$1\ 10^7$
Hg-193m	$1\ 10^1$	$1\ 10^6$	$1\ 10^{12}$	$1\ 10^7$
Hg-194 +	$1\ 10^1$	$1\ 10^6$	$1\ 10^{10}$	$1\ 10^7$
Hg-195	$1\ 10^2$	$1\ 10^6$	$1\ 10^{12}$	$1\ 10^7$
Hg-195m + (organic)	$1\ 10^2$	$1\ 10^6$	$1\ 10^{12}$	$1\ 10^7$
Hg-195m + (inorganic)	$1\ 10^2$	$1\ 10^6$	$1\ 10^{11}$	$1\ 10^7$
Hg-197	$1\ 10^2$	$1\ 10^7$	$1\ 10^{12}$	$1\ 10^8$
Hg-197m (organic)	$1\ 10^2$	$1\ 10^6$	$1\ 10^{12}$	$1\ 10^7$
Hg-197m (inorganic)	$1\ 10^2$	$1\ 10^6$	$1\ 10^{11}$	$1\ 10^7$

1	2	3	4	5
Radionuclide name, symbol, isotope	Concentration for notification. Regulation 6 and Schedule 1 (Bq/g)	Quantity for notification. Regulation 6 and Schedule 1 (Bq)	Quantity for notification of occurrences. Regulation 30(1) (Bq)	Quantity for notification of occurrences. Regulation 30(3) (Bq)
Hg-199m	$1\ 10^2$	$1\ 10^6$	$1\ 10^{14}$	$1\ 10^7$
Hg-203	$1\ 10^2$	$1\ 10^5$	$1\ 10^{11}$	$1\ 10^6$
Thallium				
Tl-194	$1\ 10^1$	$1\ 10^6$	$1\ 10^{13}$	$1\ 10^7$
Tl-194m	$1\ 10^1$	$1\ 10^6$	$1\ 10^{13}$	$1\ 10^7$
Tl-195	$1\ 10^1$	$1\ 10^6$	$1\ 10^{13}$	$1\ 10^7$
Tl-197	$1\ 10^2$	$1\ 10^6$	$1\ 10^{13}$	$1\ 10^7$
Tl-198	$1\ 10^1$	$1\ 10^6$	$1\ 10^{12}$	$1\ 10^7$
Tl-198m	$1\ 10^1$	$1\ 10^6$	$1\ 10^{13}$	$1\ 10^7$
Tl-199	$1\ 10^2$	$1\ 10^6$	$1\ 10^{13}$	$1\ 10^7$
Tl-200	$1\ 10^1$	$1\ 10^6$	$1\ 10^{11}$	$1\ 10^7$
Tl-201	$1\ 10^2$	$1\ 10^6$	$1\ 10^{12}$	$1\ 10^7$
T-202	$1\ 10^2$	$1\ 10^6$	$1\ 10^{11}$	$1\ 10^7$
Tl-204	$1\ 10^4$	$1\ 10^4$	$1\ 10^{11}$	$1\ 10^5$
Lead				
Pb-195m	$1\ 10^1$	$1\ 10^6$	$1\ 10^{13}$	$1\ 10^7$
Pb-198	$1\ 10^2$	$1\ 10^6$	$1\ 10^{13}$	$1\ 10^7$
Pb-199	$1\ 10^1$	$1\ 10^6$	$1\ 10^{13}$	$1\ 10^7$
Pb-200	$1\ 10^2$	$1\ 10^6$	$1\ 10^{12}$	$1\ 10^7$
Pb-201	$1\ 10^1$	$1\ 10^6$	$1\ 10^{12}$	$1\ 10^7$
Pb-202	$1\ 10^3$	$1\ 10^6$	$1\ 10^{10}$	$1\ 10^7$
Pb-202m	$1\ 10^1$	$1\ 10^6$	$1\ 10^{12}$	$1\ 10^7$
Pb-203	$1\ 10^2$	$1\ 10^6$	$1\ 10^{12}$	$1\ 10^7$
Pb-205	$1\ 10^4$	$1\ 10^7$	$1\ 10^{11}$	$1\ 10^8$
Pb-209	$1\ 10^5$	$1\ 10^6$	$1\ 10^{14}$	$1\ 10^7$
Pb-210+	$1\ 10^1$	$1\ 10^4$	$1\ 10^8$	$1\ 10^5$
Pb-211	$1\ 10^2$	$1\ 10^6$	$1\ 10^{12}$	$1\ 10^7$
Pb-212+	$1\ 10^1$	$1\ 10^5$	$1\ 10^{10}$	$1\ 10^6$
Pb-214	$1\ 10^2$	$1\ 10^6$	$1\ 10^{12}$	$1\ 10^7$
Bismuth				
Bi-200	$1\ 10^1$	$1\ 10^6$	$1\ 10^{13}$	$1\ 10^7$
Bi-201	$1\ 10^1$	$1\ 10^6$	$1\ 10^{12}$	$1\ 10^7$
Bi-202	$1\ 10^1$	$1\ 10^6$	$1\ 10^{12}$	$1\ 10^7$

1	2	3	4	5
Radionuclide name, symbol, isotope	Concentration for notification. Regulation 6 and Schedule 1 (Bq/g)	Quantity for notification. Regulation 6 and Schedule 1 (Bq)	Quantity for notification of occurrences. Regulation 30(1) (Bq)	Quantity for notification of occurrences. Regulation 30(3) (Bq)
Bi-203	$1\ 10^{1}$	$1\ 10^{6}$	$1\ 10^{11}$	$1\ 10^{7}$
Bi-205	$1\ 10^{1}$	$1\ 10^{6}$	$1\ 10^{11}$	$1\ 10^{7}$
Bi-206	$1\ 10^{1}$	$1\ 10^{5}$	$1\ 10^{10}$	$1\ 10^{6}$
Bi-207	$1\ 10^{1}$	$1\ 10^{6}$	$1\ 10^{10}$	$1\ 10^{7}$
Bi-210	$1\ 10^{3}$	$1\ 10^{6}$	$1\ 10^{9}$	$1\ 10^{7}$
Bi-210m+	$1\ 10^{1}$	$1\ 10^{5}$	$1\ 10^{8}$	$1\ 10^{6}$
Bi-212+	$1\ 10^{1}$	$1\ 10^{5}$	$1\ 10^{11}$	$1\ 10^{6}$
Bi-213	$1\ 10^{2}$	$1\ 10^{6}$	$1\ 10^{11}$	$1\ 10^{7}$
Bi-214	$1\ 10^{1}$	$1\ 10^{5}$	$1\ 10^{12}$	$1\ 10^{6}$
Polonium				
Po-203	$1\ 10^{1}$	$1\ 10^{6}$	$1\ 10^{13}$	$1\ 10^{7}$
Po-205	$1\ 10^{1}$	$1\ 10^{6}$	$1\ 10^{12}$	$1\ 10^{7}$
Po-206	$1\ 10^{1}$	$1\ 10^{6}$	$1\ 10^{11}$	$1\ 10^{7}$
Po-207	$1\ 10^{1}$	$1\ 10^{6}$	$1\ 10^{12}$	$1\ 10^{7}$
Po-208	$1\ 10^{1}$	$1\ 10^{4}$	$1\ 10^{7}$	$1\ 10^{5}$
Po-209	$1\ 10^{1}$	$1\ 10^{4}$	$1\ 10^{7}$	$1\ 10^{5}$
Po-210	$1\ 10^{1}$	$1\ 10^{4}$	$1\ 10^{7}$	$1\ 10^{5}$
Astatine				
At-207	$1\ 10^{1}$	$1\ 10^{6}$	$1\ 10^{12}$	$1\ 10^{7}$
At-211	$1\ 10^{3}$	$1\ 10^{7}$	$1\ 10^{10}$	$1\ 10^{8}$
Francium				
Fr-222	$1\ 10^{3}$	$1\ 10^{5}$	$1\ 10^{12}$	$1\ 10^{6}$
Fr-223	$1\ 10^{2}$	$1\ 10^{6}$	$1\ 10^{13}$	$1\ 10^{7}$
Radon				
Rn-220+	$1\ 10^{4}$	$1\ 10^{7}$	$1\ 10^{8}$	$1\ 10^{8}$
Rn-222+	$1\ 10^{1}$	$1\ 10^{8}$	$1\ 10^{9}$	$1\ 10^{9}$
Radium				
Ra-223+	$1\ 10^{2}$	$1\ 10^{5}$	$1\ 10^{7}$	$1\ 10^{6}$
Ra-224+	$1\ 10^{1}$	$1\ 10^{5}$	$1\ 10^{8}$	$1\ 10^{6}$
Ra-225	$1\ 10^{2}$	$1\ 10^{5}$	$1\ 10^{7}$	$1\ 10^{6}$
Ra-226+	$1\ 10^{1}$	$1\ 10^{4}$	$1\ 10^{7}$	$1\ 10^{5}$

1	2	3	4	5
Radionuclide name, symbol, isotope	Concentration for notification. Regulation 6 and Schedule 1 (Bq/g)	Quantity for notification. Regulation 6 and Schedule 1 (Bq)	Quantity for notification of occurrences. Regulation 30(1) (Bq)	Quantity for notification of occurrences. Regulation 30(3) (Bq)
Ra-227	$1\ 10^2$	$1\ 10^6$	$1\ 10^{13}$	$1\ 10^7$
Ra-228+	$1\ 10^1$	$1\ 10^5$	$1\ 10^8$	$1\ 10^6$
Actinium				
Ac-224	$1\ 10^2$	$1\ 10^6$	$1\ 10^{10}$	$1\ 10^7$
Ac-225+	$1\ 10^1$	$1\ 10^4$	$1\ 10^7$	$1\ 10^5$
Ac-226	$1\ 10^2$	$1\ 10^5$	$1\ 10^8$	$1\ 10^6$
Ac-227+	$1\ 10^{-1}$	$1\ 10^3$	$1\ 10^5$	$1\ 10^4$
Ac-228	$1\ 10^1$	$1\ 10^6$	$1\ 10^{10}$	$1\ 10^7$
Thorium				
Th-226+	$1\ 10^3$	$1\ 10^7$	$1\ 10^{11}$	$1\ 10^8$
Th-227	$1\ 10^1$	$1\ 10^4$	$1\ 10^7$	$1\ 10^5$
Th-228+	$1\ 10^0$	$1\ 10^4$	$1\ 10^6$	$1\ 10^5$
Th-229+	$1\ 10^0$	$1\ 10^3$	$1\ 10^6$	$1\ 10^4$
Th-230	$1\ 10^0$	$1\ 10^4$	$1\ 10^6$	$1\ 10^5$
Th-231	$1\ 10^3$	$1\ 10^7$	$1\ 10^{12}$	$1\ 10^8$
Th-232	$1\ 10^1$	$1\ 10^4$	$1\ 10^6$	$1\ 10^5$
Th-232sec	$1\ 10^0$	$1\ 10^3$	$1\ 10^6$	$1\ 10^4$
Th-234+	$1\ 10^3$	$1\ 10^5$	$1\ 10^{10}$	$1\ 10^6$
Protactinium				
Pa-227	$1\ 10^3$	$1\ 10^6$	$1\ 10^{11}$	$1\ 10^7$
Pa-228	$1\ 10^1$	$1\ 10^6$	$1\ 10^{10}$	$1\ 10^7$
Pa-230	$1\ 10^1$	$1\ 10^6$	$1\ 10^8$	$1\ 10^7$
Pa-231	$1\ 10^0$	$1\ 10^3$	$1\ 10^6$	$1\ 10^4$
Pa-232	$1\ 10^1$	$1\ 10^6$	$1\ 10^{10}$	$1\ 10^7$
Pa-233	$1\ 10^2$	$1\ 10^7$	$1\ 10^{10}$	$1\ 10^8$
Pa-234	$1\ 10^1$	$1\ 10^6$	$1\ 10^{12}$	$1\ 10^7$
Uranium				
U-230+	$1\ 10^1$	$1\ 10^5$	$1\ 10^7$	$1\ 10^6$
U-231	$1\ 10^2$	$1\ 10^7$	$1\ 10^{11}$	$1\ 10^8$
U-232+	$1\ 10^0$	$1\ 10^3$	$1\ 10^6$	$1\ 10^4$
U-233	$1\ 10^1$	$1\ 10^4$	$1\ 10^7$	$1\ 10^5$
U-234	$1\ 10^1$	$1\ 10^4$	$1\ 10^7$	$1\ 10^5$

1	2	3	4	5
Radionuclide name, symbol, isotope	Concentration for notification. Regulation 6 and Schedule 1 (Bq/g)	Quantity for notification. Regulation 6 and Schedule 1 (Bq)	Quantity for notification of occurrences. Regulation 30(1) (Bq)	Quantity for notification of occurrences. Regulation 30(3) (Bq)
U-235 +	$1\ 10^1$	$1\ 10^4$	$1\ 10^7$	$1\ 10^5$
U-236	$1\ 10^1$	$1\ 10^4$	$1\ 10^7$	$1\ 10^5$
U-237	$1\ 10^2$	$1\ 10^6$	$1\ 10^{11}$	$1\ 10^7$
U-238 +	$1\ 10^1$	$1\ 10^4$	$1\ 10^7$	$1\ 10^5$
U-238 sec	$1\ 10^0$	$1\ 10^3$	$1\ 10^6$	$1\ 10^4$
U-239	$1\ 10^2$	$1\ 10^6$	$1\ 10^{14}$	$1\ 10^7$
U-240	$1\ 10^3$	$1\ 10^7$	$1\ 10^{12}$	$1\ 10^8$
U-240 +	$1\ 10^1$	$1\ 10^6$	$1\ 10^{11}$	$1\ 10^7$
Neptunium				
Np-232	$1\ 10^1$	$1\ 10^6$	$1\ 10^{13}$	$1\ 10^7$
Np-233	$1\ 10^2$	$1\ 10^7$	$1\ 10^{14}$	$1\ 10^8$
Np-234	$1\ 10^1$	$1\ 10^6$	$1\ 10^{11}$	$1\ 10^7$
Np-235	$1\ 10^3$	$1\ 10^7$	$1\ 10^{11}$	$1\ 10^8$
Np-236 ($1.15\ 10^5$ years)	$1\ 10^2$	$1\ 10^5$	$1\ 10^8$	$1\ 10^6$
Np-236 (22.5 hours)	$1\ 10^3$	$1\ 10^7$	$1\ 10^{11}$	$1\ 10^8$
Np-237 +	$1\ 10^0$	$1\ 10^3$	$1\ 10^7$	$1\ 10^4$
Np-238	$1\ 10^2$	$1\ 10^6$	$1\ 10^{11}$	$1\ 10^7$
Np-239	$1\ 10^2$	$1\ 10^7$	$1\ 10^{11}$	$1\ 10^8$
Np-240	$1\ 10^1$	$1\ 10^6$	$1\ 10^{13}$	$1\ 10^7$
Plutonium				
Pu-234	$1\ 10^2$	$1\ 10^7$	$1\ 10^{10}$	$1\ 10^8$
Pu-235	$1\ 10^2$	$1\ 10^7$	$1\ 10^{14}$	$1\ 10^8$
Pu-236	$1\ 10^1$	$1\ 10^4$	$1\ 10^7$	$1\ 10^5$
Pu-237	$1\ 10^3$	$1\ 10^7$	$1\ 10^{11}$	$1\ 10^8$
Pu-238	$1\ 10^0$	$1\ 10^4$	$1\ 10^6$	$1\ 10^5$
Pu-239	$1\ 10^0$	$1\ 10^4$	$1\ 10^6$	$1\ 10^5$
Pu-240	$1\ 10^0$	$1\ 10^3$	$1\ 10^6$	$1\ 10^4$
Pu-241	$1\ 10^2$	$1\ 10^5$	$1\ 10^8$	$1\ 10^6$
Pu-242	$1\ 10^0$	$1\ 10^4$	$1\ 10^6$	$1\ 10^5$
Pu-243	$1\ 10^3$	$1\ 10^7$	$1\ 10^{13}$	$1\ 10^8$
Pu-244	$1\ 10^0$	$1\ 10^4$	$1\ 10^6$	$1\ 10^5$
Pu-245	$1\ 10^2$	$1\ 10^6$	$1\ 10^{12}$	$1\ 10^7$
Pu-246	$1\ 10^2$	$1\ 10^6$	$1\ 10^{10}$	$1\ 10^7$

1	2	3	4	5
Radionuclide name, symbol, isotope	Concentration for notification. Regulation 6 and Schedule 1 (Bq/g)	Quantity for notification. Regulation 6 and Schedule 1 (Bq)	Quantity for notification of occurrences. Regulation 30(1) (Bq)	Quantity for notification of occurrences. Regulation 30(3) (Bq)
Americium				
Am-237	$1\ 10^2$	$1\ 10^6$	$1\ 10^{13}$	$1\ 10^7$
Am-238	$1\ 10^1$	$1\ 10^6$	$1\ 10^{13}$	$1\ 10^7$
Am-239	$1\ 10^2$	$1\ 10^6$	$1\ 10^{12}$	$1\ 10^7$
Am-240	$1\ 10^1$	$1\ 10^6$	$1\ 10^{11}$	$1\ 10^7$
Am-241	$1\ 10^0$	$1\ 10^4$	$1\ 10^6$	$1\ 10^5$
Am-242	$1\ 10^3$	$1\ 10^6$	$1\ 10^{10}$	$1\ 10^7$
Am-242m +	$1\ 10^0$	$1\ 10^4$	$1\ 10^6$	$1\ 10^5$
Am-243 +	$1\ 10^0$	$1\ 10^3$	$1\ 10^6$	$1\ 10^4$
Am-244	$1\ 10^1$	$1\ 10^6$	$1\ 10^{11}$	$1\ 10^7$
Am-244m	$1\ 10^4$	$1\ 10^7$	$1\ 10^{14}$	$1\ 10^8$
Am-245	$1\ 10^3$	$1\ 10^6$	$1\ 10^{13}$	$1\ 10^7$
Am-246	$1\ 10^1$	$1\ 10^5$	$1\ 10^{13}$	$1\ 10^6$
Am-246m	$1\ 10^1$	$1\ 10^6$	$1\ 10^{13}$	$1\ 10^7$
Curium				
Cm-238	$1\ 10^2$	$1\ 10^7$	$1\ 10^{12}$	$1\ 10^8$
Cm-240	$1\ 10^2$	$1\ 10^5$	$1\ 10^7$	$1\ 10^6$
Cm-241	$1\ 10^2$	$1\ 10^6$	$1\ 10^9$	$1\ 10^7$
Cm-242	$1\ 10^2$	$1\ 10^5$	$1\ 10^7$	$1\ 10^6$
Cm-243	$1\ 10^0$	$1\ 10^4$	$1\ 10^7$	$1\ 10^5$
Cm-244	$1\ 10^1$	$1\ 10^4$	$1\ 10^7$	$1\ 10^5$
Cm-245	$1\ 10^0$	$1\ 10^3$	$1\ 10^6$	$1\ 10^4$
Cm-246	$1\ 10^0$	$1\ 10^3$	$1\ 10^6$	$1\ 10^4$
Cm-247	$1\ 10^0$	$1\ 10^4$	$1\ 10^6$	$1\ 10^5$
Cm-248	$1\ 10^0$	$1\ 10^3$	$1\ 10^6$	$1\ 10^4$
Cm-249	$1\ 10^3$	$1\ 10^6$	$1\ 10^{14}$	$1\ 10^7$
Cm-250	$1\ 10^{-1}$	$1\ 10^3$	$1\ 10^5$	$1\ 10^4$
Berkelium				
Bk-245	$1\ 10^2$	$1\ 10^6$	$1\ 10^{11}$	$1\ 10^7$
Bk-246	$1\ 10^1$	$1\ 10^6$	$1\ 10^{11}$	$1\ 10^7$

1	2	3	4	5
Radionuclide name, symbol, isotope	Concentration for notification. Regulation 6 and Schedule 1 (Bq/g)	Quantity for notification. Regulation 6 and Schedule 1 (Bq)	Quantity for notification of occurrences. Regulation 30(1) (Bq)	Quantity for notification of occurrences. Regulation 30(3) (Bq)
Bk-247	$1\ 10^0$	$1\ 10^4$	$1\ 10^6$	$1\ 10^5$
Bk-249	$1\ 10^3$	$1\ 10^6$	$1\ 10^9$	$1\ 10^7$
Bk-250	$1\ 10^1$	$1\ 10^6$	$1\ 10^{12}$	$1\ 10^7$
Californium				
Cf-244	$1\ 10^4$	$1\ 10^7$	$1\ 10^{12}$	$1\ 10^8$
Cf-246	$1\ 10^3$	$1\ 10^6$	$1\ 10^9$	$1\ 10^7$
Cf-248	$1\ 10^1$	$1\ 10^4$	$1\ 10^7$	$1\ 10^5$
Cf-249	$1\ 10^0$	$1\ 10^3$	$1\ 10^6$	$1\ 10^4$
Cf-250	$1\ 10^1$	$1\ 10^4$	$1\ 10^6$	$1\ 10^5$
Cf-251	$1\ 10^0$	$1\ 10^3$	$1\ 10^6$	$1\ 10^4$
Cf-252	$1\ 10^1$	$1\ 10^4$	$1\ 10^7$	$1\ 10^5$
Cf-253	$1\ 10^2$	$1\ 10^5$	$1\ 10^8$	$1\ 10^6$
Cf-254	$1\ 10^0$	$1\ 10^3$	$1\ 10^7$	$1\ 10^4$
Einsteinium				
Es-250	$1\ 10^2$	$1\ 10^6$	$1\ 10^{13}$	$1\ 10^7$
Es-251	$1\ 10^2$	$1\ 10^7$	$1\ 10^{11}$	$1\ 10^8$
Es-253	$1\ 10^2$	$1\ 10^5$	$1\ 10^8$	$1\ 10^6$
Es-254	$1\ 10^1$	$1\ 10^4$	$1\ 10^7$	$1\ 10^5$
Es-254m	$1\ 10^2$	$1\ 10^6$	$1\ 10^9$	$1\ 10^7$
Fermium				
Fm-252	$1\ 10^3$	$1\ 10^6$	$1\ 10^9$	$1\ 10^7$
Fm-253	$1\ 10^2$	$1\ 10^6$	$1\ 10^9$	$1\ 10^7$
Fm-254	$1\ 10^4$	$1\ 10^7$	$1\ 10^{10}$	$1\ 10^8$
Fm-255	$1\ 10^3$	$1\ 10^6$	$1\ 10^9$	$1\ 10^7$
Fm-257	$1\ 10^1$	$1\ 10^5$	$1\ 10^7$	$1\ 10^6$

1	2	3	4	5
Radionuclide name, symbol, isotope	**Concentration for notification. Regulation 6 and Schedule 1 (Bq/g)**	**Quantity for notification. Regulation 6 and Schedule 1 (Bq)**	**Quantity for notification of occurrences. Regulation 30(1) (Bq)**	**Quantity for notification of occurrences. Regulation 30(3) (Bq)**
Mendelevium				
Md-257	$1\ 10^2$	$1\ 10^7$	$1\ 10^{11}$	$1\ 10^8$
Md-258	$1\ 10^2$	$1\ 10^5$	$1\ 10^7$	$1\ 10^6$
Other radionuclides not listed above (see note 1)	$1\ 10^{-1}$	$1\ 10^3$	$1\ 10^5$	$1\ 10^4$

Note 1
In the case of radionuclides not specified elsewhere in this Part, the quantities specified in this entry are to be used unless the Executive has approved some other quantity for that radionuclide.

Note 2
Nuclides carrying the suffix " + " or "sec" in the above table represent parent nuclides in equilibrium with their correspondent daughter nuclides as listed in the following Table. In this case the concentrations and quantities given in the above Table refer to the parent nuclide alone, but already take account of the daughter nuclide(s) present.

List of nuclides in secular equilibrium as referred to in note 2 of this Schedule.

Parent nuclide	Daughter nuclides
Mg-28+	Al-28
Ti-44+	Sc-44
Fe-60+	Co-60m
Ge-68+	Ga-68
Sr-82+	Rb-82
Rb-83+	Kr-83m
Y-87+	Sr-87m
Sr-90+	Y-90
Zr-93+	Nb-93m
Zr-97+	Nb-97
Tc-95m+	Tc-95
Ru-106+	Rh-106
Ag-108m+	Ag-108
Sn-121m+	Sn-121
Sn-126+	Sb-126m
Xe-122+	I-122
Cs-137+	Ba-137m
Ba-140+	La-140
Ce-144+	Pr-144
Pm-148m+	Pm-148
Gd-146+	Eu-146
Hf-172+	Lu-172
W-178+	Ta-178
W-188+	Re-188
Re-189+	Os-189m
Os-194+	Ir-194
Ir-189+	Os-189m
Pt-188+	Ir-188
Hg-194+	Au-194
Hg-195m+	Hg-195
Pb-210+	Bi-210, Po-210
Bi-210m+	Tl-206
Pb-212+	Bi-212, Tl-208, Po-212
Bi-212+	Tl-208, Po-212
Rn-220+	Po-216

Parent nuclide	Daughter nuclides

Parent nuclide	Daughter nuclides
Rn-222+	Po-218, Pb-214, Bi-214, Po-214
Ra-223+	Rn-219, Po-215, Pb-211, Bi-211, Tl-207
Ra-224+	Rn-220, Po-216, Pb-212, Bi-212, Tl-208, Po-212
Ra-226+	Rn-222, Po-218, Pb-214, Bi-214, Po-214, Pb-210, Bi-210, Po-210
Ra-228+	Ac-228
Ac-225+	Fr-221, At-217, Bi-213, Po-213, Tl-209, Pb-209
Ac-227+	Fr-223
Th-226+	Ra-222, Rn-218, Po-214
Th-228+	Ra-224, Rn-220, Po-216, Pb-212, Bi-212, Tl-208, Po-212
Th-229+	Ra-225, Ac-225, Fr-221, At-217, Bi-213, Po-213, Pb-209
Th-232sec	Ra-228, Ac-228, Th-228, Ra-224, Rn-220, Po-216, Pb-212, Bi-212, Tl-208, Po-212
Th-234+	Pa-234m
U-230+	Th-226, Ra-222, Rn-218, Po-214
U-232+	Th-228, Ra-224, Rn-220, Po-216, Pb-212, Bi-212, Tl-208, Po-212
U-235+	Th-231
U-238+	Th-234, Pa-234m
U-238sec	Th-234, Pa-234m, U-234, Th-230, Ra-226, Rn-222, Po-218, Pb-214, Bi-214, Po-214, Pb-210, Bi-210, Po-210
U-240+	Np-240
Np-237+	Pa-233
Am-242m+	Am-242
Am-243+	Np-239

PART II

QUANTITY RATIOS FOR MORE THAN ONE RADIONUCLIDE

1. For the purpose of Regulation 2(4), the quantity ratio for more than one radionuclide is the sum of the quotients of the quantity of a radionuclide present Q_p divided by the quantity of that radionuclide specified in the appropriate column of Part I of this Schedule Q_{lim}, namely—

$$\sum \frac{Q_p}{Q_{lim}}$$

2. In any case where the isotopic composition of a radioactive substance is not known or is only partially known, the quantity ratio for that substance shall be calculated by using the values specified in the appropriate column in Part I for 'other radionuclides not listed above' for any radionuclide that has not been identified or where the quantity of a radionuclide is uncertain, unless the employer can show that the use of some other value is appropriate in the circumstances of a particular case, when he may use that value.

SCHEDULE 9

Regulation 41(1)

MODIFICATIONS

The Employment Act 1989

1. In Schedule 1 to the Employment Act 1989(**a**), in place of "Parts IV and V of the Ionising Radiations Regulations 1985" there shall be substituted "Paragraphs 5 and 11 of Schedule 4 to the Ionising Radiations Regulations 1999 [S.I. 1999/xxxx]".

The Employment Rights Act 1996

2. In section 64(3) of the Employment Rights Act 1996(**b**), in place of "regulation 16 of the Ionising Radiations Regulations 1985" there shall be substituted "regulation 24 of the Ionising Radiations Regulations 1999 [S.I. 1999/xxxx]".

The Personal Protective Equipment at Work Regulations 1992

3. In regulation 3 of the Personal Protective Equipment at Work Regulations 1992(**c**), in place of "the Ionising Radiations Regulations 1985" there shall be substituted "the Ionising Radiations Regulations 1999 [S.I.1999/xxxx]".

The Notification of New Substances Regulations 1993

4. In sub-paragraph (2)(e) of regulation 3 of the Notification of New Substances Regulations 1993(**d**), in place of "the Ionising Radiations Regulations 1985" there shall be substituted "the Ionising Radiations Regulations 1999 [S.I.1999/xxxx]".

The Chemicals (Hazard Information and Packaging for Supply) Regulations 1994

5. In sub-paragraph (1)(a) of regulation 3 of the Chemicals (Hazard Information and Packaging for Supply) Regulations 1994(**e**), in place of "the Ionising Radiations Regulations 1985" there shall be substituted "the Ionising Radiations Regulations 1999 [S.I.1999/xxxx]".

The Reporting of Injuries, Diseases and Dangerous Occurrences Regulations 1995

6. Paragraph 8(2) of Schedule 2 to the Reporting of Injuries, Diseases and Dangerous Occurrences Regulations 1995(**f**), shall be deleted and the following substituted—

"In this paragraph, "radiation generator" means any electrical equipment emitting ionising radiation and containing components operating at a potential difference of more than 5kV.".

The Packaging, Labelling and Carriage of Radioactive Material by Rail Regulations 1996

7. In paragraph 1(s) of Schedule 14 to the Packaging, Labelling and Carriage of Radioactive Material by Rail Regulations 1996(**g**), after "regulation 27 of the Ionising Radiations Regulations 1985" there shall be added "or regulation 12 of the Ionising Radiations Regulations 1999 [S.1.1999/xxxx]".

(**a**) 1989 c. 38.
(**b**) 1996 c. 18.
(**c**) S.I. 1992/2966.
(**d**) S.I. 1993/3050.
(**e**) S.I. 1994/3247.
(**f**) S.I. 1995/3163.
(**g**) S.I. 1996/2090.

The Health and Safety (Enforcing Authority) Regulations 1998

8. The Health and Safety (Enforcing Authority) Regulations 1998(**a**) shall be modified as follows—
   (a) in the definition of "ionising radiation" in regulation 2(1), in place of "the Ionising Radiations Regulations 1985" there shall be substituted "the Ionising Radiations Regulations 1999 [S.I. 1999/xxxx]";
   (b) in sub-paragraph (d) of paragraph 4 of Schedule 2, in place of "Schedule 3 of the Ionising Radiations Regulations 1985" there shall be substituted "Schedule 1 of the Ionising Radiations Regulations 1999 [S.I. 1999/xxxx]".
   (c) in paragraph 5 of Schedule 2, in place of "the Ionising Radiations Regulations 1985" there shall be substituted "the Ionising Radiations Regulations 1999 [S.I.1999/xxxx]".

The Health and Safety (Fees) Regulations 1999

9. The Health and Safety (Fees) Regulations 1999(**b**) shall be modified as follows—
   (a) in sub-paragraph (1)(c) of regulation 3, in place of "the Ionising Radiations Regulations 1985" there shall be substituted "the Ionising Radiations Regulations 1999 [S.I.1999/xxxx]";
   (b) the chapeau to regulation 9 shall be deleted and the following substituted—
   "Fees for application for approval or reassessment of approval of dosimetry services and for type approval of apparatus under the Ionising Radiations Regulations 1999".
   (c) in regulation 9(2), the words "a radiation generator or an apparatus containing a radioactive substance" shall be deleted and substituted by the following—
   "apparatus pursuant to sub-paragraphs 1(c)(i) and 1(d)(i) of Schedule 1 to the Ionising Radiations Regulations 1999 [S.I.1999/xxxx]";
   (d) the title of Schedule 8 shall be deleted and the following substituted—
   "FEES FOR APPLICATIONS FOR APPROVAL OR REASSESSMENT OF APPROVAL OF DOSIMETRY SERVICES AND FOR TYPE APPROVAL OF APPARATUS UNDER THE IONISING RADIATIONS REGULATIONS 1999";
   (e) in the first entry of column 1 of Schedule 8, in place of "regulation 15 of the Ionising Radiations Regulations 1985", there shall be substituted "regulation 35 of the Ionising Radiations Regulations 1999 [S.I.1999/xxxx]";
   (f) the final entry of column 1 of Schedule 8 shall be deleted and the following substituted—
   "Type approval of apparatus under sub-paragraph 1(c)(i) or 1(d)(i) of Schedule 1 to the Ionising Radiations Regulations 1999 [S.I.1999/xxxx] (which excepts such type approved apparatus from the notification requirements of regulation 6 of those Regulations).".

(**a**) S.I. 1998/494.
(**b**) S.I. 1999/645.

EXPLANATORY NOTE

(This note is not part of the Regulations)

These Regulations supersede and consolidate the Ionising Radiations Regulations 1985 and the Ionising Radiations (Outside Workers) Regulations 1993.

The Regulations impose duties on employers to protect employees and other persons against ionising radiation arising from work with radioactive substances and other sources of ionising radiation and also impose certain duties on employees.

The Regulations implement in part as respects Great Britain provisions of—

    (a) Council Directive 96/29/Euratom (OJ No. L159, 29.6.96, p.1) laying down basic safety standards for the protection of the health of workers and the general public against the dangers arising from ionising radiation;

    (b) Council Directive 90/641/Euratom (OJ No. L349, 13.12.90, p.23) on the operational protection of outside workers exposed to the risk of ionising radiation during their activities in controlled areas;

    (c) Council Directive 97/43/Euratom (OJ No. L180, 9.7.97, p.22) on health protection of individuals against the dangers of ionising radiation in relation to medical exposure.

The Regulations are divided into 7 Parts.

Part I (Interpretation and General—Regulations 1–4)

The Regulations define the terms used in and the scope of the Regulations. For the purposes of the Regulations, an employer includes a self-employed person and an employee includes a self-employed person and a trainee.

Part II (General principles and procedures—Regulations 5–12)

The Regulations—

    (a) prohibit the carrying out of specified practices without the authorisation of the Health and Safety Executive ("the Executive");

    (b) require specified work with ionising radiation to be notified to the Executive;

    (c) require radiation employers to make a prior assessment of the risks arising from their work with ionising radiation, to make an assessment of the hazards likely to arise from that work and to prevent and limit the consequences of identifiable radiation accidents;

    (d) require radiation employers to take all necessary steps to restrict so far as is reasonably practicable the extent to which employees and other persons are exposed to ionising radiation;

    (e) require respiratory protective equipment used in work with ionising radiation to conform with agreed standards and require all personal protective equipment and other controls to be regularly examined and properly maintained;

    (f) impose limits (specified in Schedule 4) on the doses of ionising radiation which employees and other persons may receive;

    (g) require in certain circumstances the preparation of contingency plans for radiation accidents which are reasonably foreseeable.

Part III (Arrangements for the management of radiation protection—Regulations 13–15)

The Regulations require that radiation employers consult radiation protection advisers in respect of matters specified in Schedule 5 and that employers ensure that adequate information, instruction and training is given to employees and other persons. Employers are required to co-operate by exchanging information to enable compliance by others with requirements to limit the exposure of employees to ionising radiation.

Part IV (Designated areas—Regulations 16–19)

The Regulations—

    (a) provide that areas in which persons need to follow special procedures to restrict exposure or in which persons are likely to receive more than specified doses of ionising radiation be designated as controlled or supervised areas;

(b) restrict entry into controlled areas to specified persons and circumstances;

(c) require radiation employers to set out appropriate local rules for controlled or supervised areas and to appoint radiation protection supervisors for the purpose of securing compliance with the Regulations;

(d) impose specified duties upon employers in relation to outside workers;

(e) require radiation levels to be monitored in controlled or supervised areas and provide for the maintenance and testing of monitoring equipment.

Part V (Classification and monitoring of persons—Regulations 20–26)

The Regulations require that employees who are likely to receive more than specified doses of ionising radiation be designated as classified persons, that doses received by classified persons be assessed by one or more dosimetry services approved by the Executive and that records of such doses are made and kept for each such person.

The Regulations also provide for—

(a) certain employees to be subject to medical surveillance;

(b) any cases in which an employee has received an overexposure to be investigated and notified to the Executive;

(c) investigations to be made where employees are exposed above specified levels;

(d) modified dose limits for employees who have received an overexposure.

Part VI (Arrangements for the control of radioactive substances, articles and equipment—Regulations 27–33)

The Regulations—

(a) require that where a radioactive substance is to be used as a source of ionising radiation, it should, whenever reasonably practicable, be in the form of a sealed source and that any articles embodying or containing radioactive substances are suitably designed, constructed, maintained and tested;

(b) cover the accounting for, keeping and moving of radioactive substances and require that incidents in which more than specified quantities of radioactive substances escape or are lost or stolen be notified to the Executive;

(c) impose duties on manufacturers etc. and installers of articles for use in work with ionising radiation to ensure that such articles are designed, constructed and installed so as to restrict, so far as is reasonably practicable, exposure to ionising radiation;

(d) impose similar duties upon employers in relation to equipment used for medical exposures together with additional duties in relation to the testing and safe operation of such equipment;

(e) require employers to investigate any defect in medical equipment which may have resulted in a person receiving a dose of ionising radiation much greater than was intended and to notify the Executive of such incidents;

(f) prohibit interference with sources of ionising radiation.

Part VII (Duties of employees and miscellaneous—Regulations 34–41)

The Regulations impose duties upon employees engaged in carrying out work with ionising radiation. The Regulations also—

(a) provide for the approval of dosimetry services by the Executive;

(b) provide for a defence on contravention of certain regulations;

(c) provide for exemptions to be granted by the Executive;

(d) extend the provision of the Regulations outside Great Britain;

(e) contain transitional provisions; and

(f) introduce modifications relating to the Ministry of Defence.

The Regulations make consequential amendments to the enactments specified in Schedule 9 and revoke (with savings) the Ionising Radiations Regulations 1985 and the Ionising Radiations (Outside Workers) Regulations 1993.

A copy of the regulatory impact assessment prepared in respect of these Regulations can be obtained from the Health and Safety Executive, Economic Adviser's Unit, Rose Court, 2 Southwark Bridge, London SE1 9HS. A copy has been placed in the Library of each House of Parliament.

© Crown copyright 1999

Printed and published in the UK by The Stationery Office Limited
under the authority and superintendence of Carol Tullo, Controller of
Her Majesty's Stationery Office and Queen's Printer of Acts of Parliament.
WO 5847   12/99   462071   19585